Bedlam

St. Mary of Bethlehem

Terry Trainor

Copyright 2010

Preface

Fields, long green grass and the leaves on the trees are stiff with crisp rime gifted from natures faint breath. The afternoon brings rain and soaks the land. Sheep wander the wastelands their eyes peeled, looking for a quiet spot to convert into motherhood. Standing on my own where the shadows are grey watching children tend to the sheep with kind faces and gentle hands. The green fields of Bishopsgate were a picture of beauty in the thirteenth centuary in the cleansing rains and pulsing bouncing streams. White mists cling to to field and hedrow when the veil's opens the greeness makes all glad. Groups of stone brakers made their way into the wastelands in 1245 they cleared a big patch of wasteland for the building of a new priory. Groups of bullocks that had roamed these lands did not want to move. The building was to be a priory called St Mary of Bethlehem Originally opened in 1247 by Simon FitzMary, he was an Alderman and Sheriff of London. A stonebraker told his story to Simon Fitz Mary one of the sheriffs in London. 'They are not patient but will go most unwillingly with lowered head and a fuetive sideways motion. It can be seen by looking closely at them that their eyes show great horror and great fear'. Simon Fitz Mary, was one of the sheriffs of London. He made over his land west of Bishopsgate Street, near the church of St. Botolph without Bishopsgate, to Godfrey, bishop of Bethlehem. This was to build a priory of canons, brothers and sisters, of the order of St. Mary of Bethlehem. Their duties were to say prayers for the souls of the founder, of Guy de Marlowe, John Durant, Ralph Ashwye and others, a reception of the bishop of Bethlehem, and the canons and messengers of that church when they came to London. The house was to be subject to the bishop of Bethlehem. He was to receive from it an annual pension of a mark, to be increased as its wealth grew. He also had the right of visitation and correction. Fitz Mary also provided that the members of the house should wear on their copes and mantles the distinguishing sign of the order, a star, according to Matthew Paris, red with five rays inclosing a circle of blue. It doesn't seem to have lasted long as a priory, and from about 1330 onwards it is referred to as a hospital. In 1547 it was made a 'royal' hospital.

The story goes that it was first used as a hospital for the medically insane in 1403, when patients housed in a Charing Cross residence were forced to find new accommodation because the then King of England, Henry IV (1399-1413) didn't like them living so close to his palace.

Bedlam, the Beginning

The Church of the Holy Sepulchre, also called the Basilica of the Holy Sepulchre, or the Church of the Resurrection by Eastern Christians, is a church within the Christian Quarter of the walled Old City of Jerusalem. It is a few steps away from the Muristan. The site is venerated as Golgotha (the Hill of Calvary), where Jesus was crucified, and is said also to contain the place where Jesus was buried (the Sepulchre). The church has been a paramount and for many Christians the most important pilgrimage destination since at least the 4th century, as the purported site of the resurrection of Jesus. Today it also serves as the headquarters of the Greek Orthodox Patriarch of Jerusalem, while control of the building is shared between several Christian churches and secular entities in complicated arrangements essentially unchanged for centuries. Today, the church is home to Eastern Orthodoxy, Oriental Orthodoxy and Roman Catholicism. Anglican, Nontrinitarian and Protestant Christians have no permanent presence in the church and tend to venerate the alternative Garden Tomb, elsewhere in Jerusalem, as the true place of Jesus's crucifixion and resurrection. A trend among visitors to the spot (standing outside the Church) is to applaud loudly during the ringing of bells. This is to recognize the unique beauty of the Church and its unique history. The origin of this practice is in dispute; one source describes a similar act in the Peter Greenaway movie The Belly of an Architect. In the early 2nd century, the site of the present Church had been a temple of Aphrodite; several ancient writers alternatively describe it as a temple to Venus, the Roman equivalent to Aphrodite. Eusebius claims, in his Life of Constantine, that the site of the Church had originally been a Christian place of veneration, but that Hadrian had deliberately covered these Christian sites with earth, and built his own temple on top, due to his hatred for

Christianity. Although Eusebius does not say as much, the temple of Aphrodite was probably built as part of Hadrian's reconstruction of Jerusalem as Aelia Capitolina in 135, following the destruction of the Jewish Revolt of 70 and the Bar Kokhba revolt of 132–135. Emperor Constantine I ordered in about 325/326 that the temple be demolished and the soil - which had provided a flat surface for the temple - be removed, instructing Macarius of Jerusalem, the local Bishop, to build a church on the site. The Pilgrim of Bordeaux reports in 333: There, at present, by the command of the Emperor Constantine, has been built a basilica, that is to say, a church of wondrous beauty. Constantine directed his mother, Helena, to build churches upon sites which commemorated the life of Jesus Christ; she was present in 326 at the construction of the church on the site, and involved herself in the excavations and construction. During the excavation, Helena is alleged to have rediscovered the True Cross, and a tomb, though Eusebius's account makes no mention of Helena's presence at the excavation, nor of the finding of the cross but only the tomb. According to Eusebius, the tomb exhibited a clear and visible proof that it was the tomb of Jesus. Socrates Scholasticus (born c. 380), in his Ecclesiastical History, gives a full description of the discovery[9] (that was repeated later by Sozomen and by Theodoret) which emphasizes the role played in the excavations and construction by Helena; just as the Church of the Nativity in Bethlehem (also founded by Constantine and Helena) commemorated the birth of Jesus, the Church of the Holy Sepulchre would commemorate his death and resurrection. Constantine's church was built as two connected churches over the two different holy sites, including a great basilica (the Martyrium visited by Egeria in the 380s), an enclosed colonnaded atrium (the Triportico) with the traditional site of Golgotha in one corner, and a rotunda, called the Anastasis ("Resurrection"), which contained the remains of a rock-cut room that Helena and Macarius identified as the burial site of Jesus. The rockface at the west end of the building was cut away, although it is unclear how much remained in Constantine's time, as archaeological investigation has revealed that the temple of Aphrodite reached far into the current rotunda area, and the temple enclosure would therefore have reached even further to the west. According to tradition, Constantine arranged for the rockface to be removed from around the tomb, without harming it, in order to isolate the tomb; in the centre of the rotunda is a small building called the Kouvouklion (Κουβούκλιον; Modern Greek for small compartment) or Aedicule (from Latin: aediculum, small building), which supposedly encloses this tomb, although it is not currently possible to verify the claim, as the remains are completely enveloped by a marble sheath. The discovery of the kokhim tombs just beyond the west end of the Church, and more recent archaeological investigation of the rotunda floor, suggest that a narrow spur of at least ten yards length would have had to jut out from the rock face if the contents of the Aedicule were once inside it. The dome of the rotunda was completed by the end of the 4th century. Each year, the Eastern Orthodox Church celebrates the anniversary of the consecration of the Church of the Resurrection (Holy Sepulchre) on September 13 (for those churches which follow the traditional Julian calendar, September 13 currently falls on September 26 of the modern Gregorian calendar). This building was damaged by fire in 614 when the Persians, under Khosrau II, invaded Jerusalem and captured the Cross. In 630, Emperor Heraclius marched triumphantly into Jerusalem and restored the True Cross to the rebuilt Church of the Holy Sepulchre. Under the Muslims it remained a Christian church. The early Muslim rulers protected the city's Christian sites, prohibiting their destruction and their use as living quarters. In 966 the doors and roof were burnt during a riot. On October 18, 1009, Fatimid caliph Al-Hakim bi-Amr Allah ordered the complete destruction of the church. It is believed that Al-Hakim "was aggrieved by the scale of the Easter pilgrimage to Jerusalem, which was caused specially by the annual miracle of the Holy Fire within the Sepulchre.[13] The measures against the church were part of a more general campaign against Christian places of worship in Palestine and Egypt, which involved a great deal of other damage: Adhemar of Chabannes recorded that the church of St George at Lydda 'with many other churches of the saints' had been attacked, and the 'basilica of the Lord's Sepulchre destroyed down to the ground'. ...The Christian writer Yahya ibn Sa'id reported that everything was razed 'except those parts which were impossible to destroy or would have been too difficult to carry away'." The Church's foundations were hacked down to bedrock. The Edicule and the east and west walls and the roof of the cut-rock tomb it encased were destroyed or damaged (contemporary accounts vary), but the north and south walls were likely protected by rubble from further damage. The "mighty pillars resisted destruction up to the height of the gallery pavement, and are now effectively the only remnant of the fourth-century buildings." Some minor repairs were done to the section believed to be the tomb of Jesus almost immediately after 1009, but a true attempt at restoration would have to wait for decades. European reaction was of shock and dismay, with far-reaching and intense consequences. For example, Cluniac monk Rodulfus Glaber blamed the Jews, with the result that Jews were expelled from Limoges and other French towns. Ultimately, this destruction provided an impetus to the later Crusades.

Reconstruction

In wide ranging negotiations between the Fatimids and the Byzantine Empire in 1027-8 an agreement was reached whereby the new Caliph Ali az-Zahir (Al-Hakim's son) agreed to allowing the rebuilding and redecoration of the Church. The rebuilding was finally completed with the financing of the huge expense by Emperor Constantine IX Monomachos and Patriarch Nicephorus of Constantinople in 1048. As a concession, the mosque in Constantinople was re-opened and sermons were to be pronounced in az-Zahir's name. Muslim sources say a by-product of the agreement was the recanting of Islam by many Christians who had been forced to convert under Al-Hakim's persecutions. In addition the Byzantines, while releasing 5,000 Muslim prisoners, made demands for the restoration of other churches destroyed by Al-Hakim and the re-establishment of a Patriarch in Jerusalem. Contemporary sources credit the emperor with spending vast sums in an effort to restore the Church of the Holy Sepulchre after this agreement was made. Despite the Byzantines spending vast sums on the project, "a total replacement was far beyond available resources. The new construction was concentrated on the rotunda and its surrounding buildings: the great basilica remained in ruins." The rebuilt church site consisted of "a court open to the sky, with five small chapels attached to it." The chapels were "to the east of the court of resurrection, where the wall of the great church had been. They commemorated scenes from the passion, such as the location of the prison of Christ and of his flagellation, and presumably were so placed because of the difficulties for free movement among shrines in the streets of the city. The dedication of these chapels indicates the importance of the pilgrims' devotion to the suffering of Christ. They have been described as 'a sort of Via Dolorosa in miniature'... since little or no rebuilding took place on the site of the great basilica. Western pilgrims to Jerusalem during the eleventh century found much of the sacred site in ruins." Control of Jerusalem, and thereby the Church of the Holy Sepulchre, continued to change hands several times between the Fatimids and the Seljuk Turks (loyal to the Abbasid caliph in Baghdad) until the arrival of the Crusaders in 1099. Many historians maintain that the main concern of Pope Urban II, when calling for the First Crusade, was the threat to Constantinople from the Turkish invasion of Asia Minor in response to the appeal of Emperor Alexios I Komnenos Historians agree that the fate of Jerusalem and thereby the Church of the Holy Sepulchre was of concern if not the immediate goal of papal policy in 1095. The idea of taking Jerusalem gained more focus as the Crusade was underway. The rebuilt church site was taken from the Fatimids (who had recently taken it from the Abassids) by the knights of the First Crusade on 15 July 1099. The First Crusade was envisioned as an armed pilgrimage, and no crusader could consider his journey complete unless he had prayed as a pilgrim at the Holy Sepulchre. Crusader Prince Godfrey of Bouillon, who became the first crusader monarch of Jerusalem, decided not to use the title "king" during his lifetime, and declared himself Advocatus Sancti Sepulchri (Protector (or Defender) of the Holy Sepulchre). By the crusader period, a cistern under the former basilica was rumoured to have been the location that Helena had found the True Cross, and began to be venerated as such; although the cistern later became the Chapel of the Invention of the Cross, there is no evidence for the rumour prior to the 11th century, and modern archaeological investigation has now dated the cistern to the 11th century repairs by Monomachos. The chronicler William of Tyre reports on the renovation of the Church in the mid-12th century. The crusaders investigated the eastern ruins on the site, occasionally excavating through the rubble, and while attempting to reach the cistern, they discovered part of the original ground level of Hadrian's temple enclosure; they decided to transform this space into a chapel dedicated to Helena (the Chapel of Saint Helena), widening their original excavation tunnel into a proper staircase. The crusaders began to refurnish the church in a Romanesque style and added a bell tower. These renovations unified the small chapels on the site and were completed during the reign of Queen Melisende in 1149, placing all the Holy places under one roof for the first time. The church became the seat of the first Latin Patriarchs, and was also the site of the kingdom's scriptorium. The church was lost to Saladin, along with the rest of the city, in 1187, although the treaty established after the Third Crusade allowed for Christian pilgrims to visit the site. Emperor Frederick II regained the city and the church by treaty in the 13th century, while he himself was under a ban of excommunication, leading to the curious result of the holiest church in Christianity being laid under interdict. Both city and church were captured by the Khwarezmians in 1244.

The Mendicant Friars 1216 1227

There was very little the Church could do to attack the underlying problems that had given rise to the popular heresies of the twelfth century. Innocent III and his immediate successors attacked the symptoms of these problems, and used the weapons of the Inquisition and the crusade to crush anti-clericalism and heresy wherever possible. One should note that the evil reputation of the Inquisition is largely undeserved. In 16th-century Spain, the monarchs gained control of the Inquisition and used it as a thought police and as a way of attacking enemies who were guilty of no crimes under secular law. The medieval Inquisition, formally organized at the Fourth Lateran Council in 1215, was a repressive institution, but was not guilty of the excesses characteristic of the early modern period. Bishops had always had the power to question and try alleged heretics in their Episcopal courts, but the Inquisition brought this function under a single organization that developed a standard procedure and regulations. Alleged heretics were interviewed at length and, if they were found to hold beliefs contrary to the "revealed truth" taught by the Church, were instructed in correct doctrine and allowed to recant (renounce) that belief and accept the Church's teaching. They were then allowed to go free, although often required to perform heavy penance. If they were charged with having returned to their old beliefs, they were subjected to a much more intensive questioning (although torture was not employed). If it was found that they had in fact returned to their error, they could be declared heretics and excommunicated, or expelled from the community of the faithful. They were then turned over to secular authorities, and usually imprisoned or executed, the latter often being done in savage and cruel ways. Although the Inquisition was in many ways hypocritical and unjust, it was an effective tool against heretical movements. In the long run, however, it was an admission of moral failure and buttressed the Church's position by instilling fear rather than promoting faith. It was a negative solution. The rise of the mendicant friars provided the positive answer to the challenge presented by the popular heresies.

The Dominicans

Dominic (Domingo) (1170-1221) was a Navarrese who became a cleric at an early age and found himself working in Southern France at the time of the Albigensian Crusade. He was shocked by the heresies and resolved to fight them personally through preaching. After some success, he was given a building in Toulouse in which he set up a rehabilitation center for heretics and a home for the orphaned girls of southern French nobles killed in the battles of the time. He also set up a center for the education of clerics in Church doctrine, the niceties of religious debate, and the art of preaching. His following grew to the point where he appealed to Pope Honorius III (1216-1227) for official recognition. The pope granted this in 1216, and the recognition was confirmed in the Fourth lateran Council. His organization took the name of the Order of Preaching Friars and adopted the Benedictine Rule modified to meet his special aims. The Order followed the example of the Albigensians and Waldensians. Although its members took monastic vows of poverty, chastity, and obedience, they were to work in the secular world, traveling about in pairs, preaching in the vernacular, and -- in order to avoid the suspicion that they represented the wealth and power of the Church -- to beg their food from the laity. They were also expected to be learned, and, to achieve this end, Dominic began to set up training centers and bases from which his followers could operate. The movement proved quite attractive to men with high ideals, and, by 1221, there were some sixty Dominican centers in operation. Since Dominic realized that it was necessary for the Church to keep the loyalty of the educated classes, Dominicans soon began working within the new universities, as students themselves, as masters offering their own courses of study, attracting students, setting up student hospices, establishing the equivalent of scholarships, and training the next generation of faculty. Quite soon, Dominicans such as Albertus Magnus and Thomas Aquinas were gaining the Order considerable respect among the educated and intellectuals of western Europe. To sum up, the Dominicans followed an ascetic way of life that did much to allay middle class suspicions of the Church's apparent preoccupation with wealth and ostentation. They made an effective appeal to the intelligentsia, thus strengthening their own Order and strengthening the loyalty of the educated classes. Their evangelism was quite effective, and they set a pattern which increased the popularity of the sermon as a tool of religious instruction and broadened the use of confession in focusing the Church's attention on the needs of the individual. Their concern for the common people emphasized the social functions of the Church at a time when those functions needed greater attention. The Dominicans eventually came to provide the personnel for the Inquisition, and contributed a great deal to the discretion and humanity that characterized that institution in its early days. As they grew more and more successful, however, the laity began to endow the Order with more property and wealth. Although they tried to separate themselves from the management of these possessions, by the end of the thirteenth century, they were suffering from their own success, had become wealthy, and were attracting new members more impressed by the Order's wealth and prestige than its original aims. This was perhaps a result of the tendency of the Dominicans to consider their practices as means to an end rather than a good in and of themselves. Consequently, they failed to develop the intense and all-encompassing ideal of Christian action that the times required. This achievement was reserved for the Franciscans.

The Franciscans

It would not be too sweeping a statement to say that Francis of Assisi (1182-1226) embodied the true religious aspirations of the men and women of thirteenth- century Europe or that he has become the most beloved figure of the entire medieval period. It is important to realize, however, that he was also a revolutionary figure and that the Church was hard-pressed to contain and control the social forces that he inspired. He was both beloved and quite dangerous. Francis was born in the north Italian hill-town of Assisi, the indulged son of a rich silk merchant. He led a more or less wild life, taking the troubadours and chivalric nobles as his ideals. At the age of twenty, he left on a military expedition, but fell ill and had to return home. After recovering, he threw all of his friends a raucous banquet and, after considerable drinking, led them in an impromptu parade through the streets of the town. When his friends found that he was missing, they retraced their steps and found him deep in a trance. He had undergone that sudden and intense inner conversion that the men and women of the period called "religion." He began to spend money lavishly in charity to the poor, so much so that his father took steps to disown him before he could bankrupt the family. Francis responded by stripping himself naked, giving everything that he had back to his father and going into the forest to live as a hermit in a hut of twigs. He was soon joined by some of his young drinking- companions and began wondering what God wanted him to do. In a divination practice common in the period, he opened a Bible three times to a passage chosen at random. Each time, his finger lit on the passage "Give all that thou hast to the poor, and follow me." He and his friends decided that they were supposed to pattern their lives on Jess and his disciples. After a while, they began to go out from their mountainside wilderness in pairs like the Albigensians and Waldensians, preaching and practicing acts of charity. They resolved to own nothing and the beg jobs in return for their daily meals, making absolutely no provision for the morrow. They soon found that this abandonment of secular concerns had given them a great sense of freedom and began experiencing ecstatic trances and mystic experiences. The pattern of the Franciscan movement, as embodied in Francis himself, took shape during this period.

Francis's Ideals

Service to humanity particularly the poor and helpless were granted. This was, curiously enough, combined with the ideal of the crusade. Francis believed that they should seek to convert the Muslims rather than simply fighting them. Francis was uninterested in intellectual pursuits. He felt that religion was a matter of the heart, not of the mind. He was imbued with a romantic ideal, considering himself a troubadour of God and a wooer of Our Lady Poverty. He also felt himself to be a part of the natural world, a startling break with the past tradition of viewing Nature as an enemy to be subdued. He did not view humility as an exercise to subdue the sense of self, but was humble because he felt himself to be humble. Humility was, for Francis and his followers, a recognition and acceptance of one's self. He was a practical mystic, emphasizing the need for personal, direct and individual union with God. His sense of piety was a natural one. His obedience to the Church was based upon his own lack of interest in theological matters. His sense of piety was a natural one. His obedience to the Church was based upon his own lack of interest in theological matters. He and his followers were filled with a sense of personal joy that was evident to all who crossed their paths.

Growth of the Franciscans

In 1209, Francis went to Pope Innocent III (1198-1216) to appeal for recognition. Innocent doubted that Francis and his followers could follow the life of poverty they had set out for themselves, but was perhaps unaware that many people in Europe lived in exactly this fashion. At any rate, he gave a somewhat vague and oral permission for them to pursue what they had proposed. The movement gained force rapidly, and was recognized in 1217 in the Fourth Lateran Council as the Friars Minor, "little brothers," or "lesser brothers," perhaps to distinguish them from the Dominicans. They were given the right to preach, and began to attract great numbers. Francis's sister, Clara, formed a women's branch, The Poor Clares, while Francis established The Third Order for people who could not become Franciscans but wished to live a life as close to that ideal as possible. In a few years, there were more than 100,000 Franciscans and at least 500,000 members of the Third Order. The reasons for this growth were several. Franciscans were expected to take the normal monastic vows, but did not have to pass through a novitiate (a probationary period, normally a year, before applicants were allowed to join a monastic order), and, unlike other orders, they could leave the ranks of the Franciscans whenever they wished. Then, too, many people lived lives of poverty not inferior to those of the Franciscans, and, but becoming Franciscans themselves, they not only gave those lives a sense of purpose and dignity, but doubtless enjoyed better treatment as Franciscans than they had experienced as mere homeless indigents. Nevertheless, others joined who were less attracted to the standards that Francis had established. The Order had grown so large that it needed administrators and some of these managers felt that the demands should be more moderate and that the Order should have the prestige and dignity enjoyed by the Dominicans and others. In 1220, while Francis was away in Egypt accompanying the Fifth Crusade, some of these administrators took control of the movement and began to establish regulations that would have changed it into something more like the Dominicans. Francis hurried back, but was able to save the situation only by agreeing to accept the Church's direction in rewriting the simple Rule that he had established in 1210. The first steps were that a novitiate had to be established and the right to leave the Order was abolished. A regular hierarchy was established and houses, again like those of the Dominicans, were established. University attendance and teaching were not only allowed, but encouraged. After having accepted these changes in the Rule of 1223, Francis withdrew from any position of leadership. When he died in 1226, he left the Testament, which he pleaded for the original ideals of the movement The Order soon broke into factions, the Spiritual Franciscans struggled to return the Order to its original conception, while others tried to moderate the rule of poverty. By the middle of the century, the Order was headed by John of Parma, a Spiritual who was attuned to the tendency of his fellows to return to the mysticism of the early days. He and others faced increasing pressure by Church authorities to moderate the Order's standards in many areas and to control the preaching of individual Franciscans more closely. It was a time of increasing class conflict, and Franciscans, particularly Spiritual Franciscans, were often found encouraging and supporting the lower classes against the upper classes favored by the Church. John (and others) responded with mystic tracts that foresaw the dawning of a new age in which the established order of things would be overthrown and the promise that blessed are the meek, for they shall inherit the Earth would be fulfilled. The Spirituals went too far with this and several, including John of Parma, were declared heretics. John was expelled from his position of leadership and imprisoned. The Order was turned over to Bonaventure and began a steady course toward more moderate practices and ideals in better harmony with those of the Church as a whole. It is tempting to view the early Franciscans as heroes and the Church as the betrayer of a noble ideal, but the matter is not as simple as that. The fact is that the population of Western Europe was growing more rapidly than its production of food, clothing, housing, fuel, and job opportunities. The Franciscans' voluntary embrace of poverty did nothing to solve the problem of poverty, and even the greatest degree of charitable sharing would have done nothing but reduce everyone to hunger at the same rate. Indeed, a cynic might say that the Franciscans were a feeble attempt to convince the indigent masses that poverty was fun. Nevertheless, it was a remarkable part of history. An economist once said that only a well-to-do society can afford to have charitable ideals. The Franciscan movement shows that this is not necessarily the case. More than that, it purchased the unified Church another three centuries of existence. It also gave the Western tradition an example of self-sacrifice and concern for the needy that has contributed greatly to our modern attitudes toward those who fall by the wayside.

In 1257 Henry III granted a protection to the brethren of St. Mary of New Bethlehem to dwell in London without Bishopsgate. Nearly a century later. In 1346, Letters Patent were issued under the Common Seal of the City whereby the House and Order of the Knights of the Blessed Mary of Bethlehem without Bishopsgate were, on the petition of Brother John Matheu, called "de Nortone" taken under the patronage and protection of the Mayor and Aldermen of the City. From the original deed of grant of Simon Fitz Mary set out in Stow, 1633, p.173, we find that the foundation was made subject to the Bishop of Bethlehem, and that the site extended from Bishopsgate Street to the Deep Ditch west and to the land of St. Botolph's Church south. The house seems to have been repaired or enlarged in 1361, as it received a bequest under the will of John Nasing to the new work of the church of St. Mary de Bethlem. The first reference to it as the Hospital of St. Mary Bethleem is in 1329, and Stow tells us that it was used as an Hospital for distracted people but when it was first so used does not appear, except that it was some time prior to 1403 Sometimes referred to as the "New Hospital without Bishopsgate," After the dissolution of the religious houses, in 1547, the king granted the custody and government of the hospital to the mayor and citizens. After this grant in 1569 Sir Thomas Roe, the Mayor, caused about an acre of land on the bank of the Deep Ditch belonging to the hospital to be enclosed to make a burial ground for such parishes in London as were in want of a burial ground.

The Hospital of St. Mary without Bishopsgate

The priory or hospital of St. Mary without Bishopsgate was founded on the east side of Bishopsgate Street by Walter Brown, a London citizen, and Rose his wife, on ground demised to them for that purpose by Walter son of Eildred, an alderman. Brown endowed it with other land adjoining, which extended to the City boundary and with 100s.rent from tenements in Blanchapelton, and in various London parishes, Allhallows Staining, St. Margaret Pattens, St. Peter the Little, St. Martin Ludgate, St. Sepulcher, and St. Martin Outwich. The foundation stone was laid by Walter, archdeacon of London, June, 1197, and the building was dedicated by William de Ste. Mère l'Eglise, bishop of London, 1199–1221, to the honour of God and the Blessed Virgin. The house consisted of Austin canons, whose duties were religious, and lay brothers and sisters to whom the care of the sick poor was entrusted, all being under the charge of a prior. The prior and brothers acknowledged themselves subject to the bishop of London, and promised that they would not make alienations of land without his leave, which he could not, however, refuse unless it was clear that loss to the hospital would result. His permission had also to be asked in case of vacancy before the canons proceeded to elect. The priory had only been in existence a short time when for some reason it was refounded in 1235, and the church was moved farther to the east. The all-important question of the water supply was settled at the end of 1277 by the gift to them of a spring called 'Snekockeswelle' in Stepney by John, bishop of London, who gave them leave to inclose it and bring the water by underground pipes into the hospital precincts. The original endowment must by this time have been supplemented by numerous grants, but the income of the hospital up to 1280 evidently did not keep pace with the expenditure, since at that date the priory owed £63 8s.for meat. Apparently all difficulty on this score had not vanished in 1303, for the archbishop of Canterbury, after a visitation, expressly stated that in his opinion the annual revenue of 300 marks was sufficient to maintain the accustomed number of inmates, twelve canons, five lay brothers, and seven sisters. Judging from these ordinances the administration of the priory had become rather lax. The ancient custom of allotting to the hospital a third of the convent flour supply, which the sisters afterwards distributed as needed, had been abandoned; bequests for special purposes had been diverted to other uses, and the lamps which at one time had been kept burning between the beds in the hospital had been taken away. The sisters seem to have received neither their proper portions of food nor their share of pittances, and no allowance was made to them for dress, which they appear to have provided for themselves out of the legacies left by their charges to the priory. With regard to the canons the archbishop ordered that money was not to be given to them for clothing, but that they should be provided with clothes uniform in colour and quality, and that on receiving the new they should give up the old; that those holding offices were to render full accounts before the whole convent, and that the cloistral canons and other hospital officials were not to go beyond the boundaries of the house singly or together, nor were they to ask leave of the prior to do so except for the evident utility of the priory. Their conduct indeed had not been exemplary: disobedience was not uncommon, and scandal and prejudice to the monastery had been caused by their frequenting the houses of Alice la Faleyse and Matilda wife of Thomas, who apparently lived within the precinct. That the canons were themselves not anxious for reform is shown by the fact that in 1306 they elected as prior a certain Robert de Cerne, a notoriously unfit person, and as such promptly deposed by Ralph, bishop of London. Ralph then exercised the right he had in such a case by appointing the sub-prior of St. Bartholomew's, Philip de London, whose probity he knew and who he hoped would improve both the tone of the house and the administration of its temporal affairs. Philip and the canons arranged that the deposed prior should receive a double allowance of bread, ale, and other food, 40s. per annum for his other necessaries, and a room near the infirmary, and for his servant a black loaf, a gallon of small beer, and one dish from the kitchen every day, and 5s. Annual wages, and that a companion should also be assigned to him. The bishopric being vacant in 1316 commissaries of the dean and chapter of St. Paul's visited St. Mary's and issued some injunctions. The canons at first declined to pay procreations, though it is difficult to see on what grounds, considering that when they needed to elect a prior in 1279 in similar circumstances they had tacitly acknowledged that the dean and chapter occupied the bishop's place. However, after a threat of excommunication they owned themselves wrong and paid the sum demanded, and the chapter of St. Paul's returned it to them for the use of the sick of the house. The better administration desired by the bishop appears to have been inaugurated by Prior Philip. The convent had been enriched to some extent between 1303 and 1331: in 1314 a chantry for four chaplains was erected by John Tany, one for two in 1325 by Roger de la Bere; in 1306 Edward I had given to the priory some land in Shalford and the advowsons of the church of Shalford with Bromley Chapel annexed, of 'Woghenersh,' (fn. 26) Puttenham, and 'Duntesfeld,' and leave to appropriate Shalford and Bromley and 'Woghenersh'; and in 1318 Edward II had granted the convent acquaintance from all tallages, aids, pontages, pavages, and other payments. When the king in 1341 ordered the exemption of the priory from payment of the subsidy, he certainly said that its endowment was so slender as hardly to suffice for the maintenance of the convent and the poor in the hospital. This, however, may be another way of stating that the charity dispensed there was very great, as he had good reason to know, more than one of his old servants finding an asylum there. The position occupied by the priory must have by this time attained some importance, for the prior was appointed one of the valuers of the 9th fleece, sheaf, &c., in co. Middlesex in 1340. The house was evidently the reverse of affluent towards the end of the fourteenth century. In 1394 a sum of £86 10s. 6d. was owing to St. Paul's Cathedral for obits, chantries, and rents unpaid in some cases for many years; (fn. 32) in 1399 the prior had to pawn a silver gilt censer for £10; and in 1400 it was arranged that in return for 300 marks granted to the prior and convent 'in their very great necessity for the relief of their house which was heavily burdened with debt,' they would give 12 marks annual quitrent from their possessions in certain London parishes to the chaplain of the chantry of St. John Baptist in St. James's Garlickhithe. The causes of its poverty can only be conjectured, but were probably the depreciation in the value of its lands owing to the Black Death, and repairs to the church and other buildings, since it is unlikely that they had escaped without much damage from the floods which in 1373 were said to occur there annually. The pope in 1391 granted an indulgence to those who visited and gave alms to the church and its chapels and to the hospital at Christmas, Easter, and other great festivals, and the benefit derived may have been considerable, for crowds of people flocked to the priory on the three days following Easter Sunday, doubtless attracted by the sermons preached at the Cross in the churchyard. One of the canons in 1389 obtained a papal indult to hold a secular benefice, and a similar grant was made to John Mildenhale, the prior, in 1401 the ordinances of William bishop of London, dated 20 June, 1431, do not disclose anything very much amiss. They chiefly concern the sisters, who as usual had been deprived of their due both as regards food and clothing. Some scandal had apparently been caused by their access to the convent kitchen, and the bishop ordered that a straight and enclosed way (via recta et clausa) should be made at the expense of the priory from the door of the sisters' house to the kitchen window, from which the sisters could, without hindrance, carry away their own dishes and those for the sick. To provide against their frequent visits to the pantry their allowance of bread and ale was to be given out weekly, though the good this would do is not very obvious, as they still had to go for bread and ale for the sick and candles for watching as needed. Anyone desiring to become a sister was to be admitted at a year's probation, and, if rejected, was to pay her own expenses, which otherwise were to be paid by the priory. At the admission and profession of a sister no exactions were to be made by the prior and convent; after profession the sisters were to be obedient to the prior, and were not to go beyond the bounds of the house except with the prior's leave and for the benefit of the house. The houses occupied by the sisters and by the sick were in need of repairs, which were to be done as quickly as the priory was able.

When Richard Cressall became prior, in 1484, he found that the property of the priory in London, the main source of the income of the house, had been allowed to fall into ruin, and it was no doubt a strain to provide for the necessary repairs and at the same time to keep up the charitable work of the hospital. More revenue was needed, and in April, 1509, King Henry VII, for £400, granted to the prior and convent in mortmain the priory of Bicknacre, where, at the death of the last prior, Edmund Godyng, only one canon was left. Its possessions included the manor of Bicknacre and thirty-one messages and land in Woodham Ferrers, Danbury, Norton, Steeple, Chelmsford, Mayland, Stow, East and West Hanningfield, Purleigh, Burnham, and Downham, and were estimated to be worth £40 10s. Per annum. Daily celebrations for the souls of the founder, benefactors, and King Henry VII were, by the bishop's orders, performed at Bicknacre by one of the canons of the New Hospital. The house in 1514 further obtained license to acquire in mortmain lands to the annual value of £100. There is no record of the light in which the religious changes of the time were regarded here, but the royal supremacy was acknowledged on 23 June, 1534, by the prior and eleven others, and it is unlikely that the king had any difficulty with the house, judging from the pensions granted at its suppression in 1538. The prior, William Major, received £80 a year, (fn. 49) and payment seems to have been made with regularity; the president, an official of whom there is no other mention, had £8 per annum; three other priests, £6 13s. 4d. each; and two others £7 10s. and £4 respectively; the two sisters 40s. each. The small number of brothers and sisters, and the state of the church, the roof of which fell before the end of the year, indicate either that the dissolution had been for some time foreseen or that much of the spirit of monasticism had departed. Whatever view is taken of the prior and canons there can be no doubt that good work was done in a hospital of 180 well-furnished beds, and Sir Richard Gresham, the mayor, in a letter to the king, begged that it might continue under the rule of the mayor and aldermen. It would, indeed, have been no more than just, for the hospital had not only been founded, but to a great extent endowed, by London citizens. The king, nevertheless, beyond allowing the sick already there to remain, turned a deaf ear to Gresham's request, and in April, 1540, a grant was made to Richard Moryson of the infirmary, the dormitory, the waste ground leading from the churchyard to the infirmary, the prior's garden and the convent garden within the enclosure, the stable in the prior's garden with some waste land adjoining, and the other tenements of the priory which extended into Shoreditch. The income of the priory, estimated in 1318 at over 300 marks, amounted in 1535 to £562 14s. 6½d. gross, and £504 12s. 11½d. net. Of this the sum of £277 13s. 4d. was derived from tenements in London and the suburbs, where the house had holdings in 1318 in thirty-seven parishes. It held, besides the property of Bicknacre Priory, in co. Middlesex the manor of Hickmans and lands and tenements called 'Burganes lands,' probably those possessed in 1318 in Shoreditch, Hackney, and Stepney; in co. Herts the manor of Beaumond Hall; in co. Essex the manor of Chalvedon, where land had been given by William Hobruge before 1318, the manor of Sabur or Seborow Hall, (fn. 67) evidently the lands in Mocking, Orsett, and Chadwell, held by the priory in the fourteenth century, the manor of Frerne or Fryern, which came into possession of the house about 1419, and lands in West Tilbury and Mountnessing; in co. Surrey the manor of Long Ditton, which, with the advowson of the church, had been given to the canons by William earl of Essex, the rectories and tithes of Shalford and Wonersh, and a pension from the church of Putney; in co. Cambridge lands and tenements in Whittlesea. A pension was also paid by the abbey of Bindon, co. Dorset.

A Hospital for the Insane 1377

Tom O' Bedlam" is the name of a critically acclaimed anonymous poem written circa 1600 (it can be definitely dated back to 1634) about a Bedlamite. The term "Tom O' Bedlam" was used in Early Modern Britain and later to describe beggars and vagrants who had or feigned mental illness (see also Abraham-men). They claimed, or were assumed, to have been former inmates at the Bethlem Royal Hospital (Bedlam). It was commonly thought that inmates were released with authority to make their way by begging, though this is probably untrue. If it happened at all the numbers were certainly small, though there were probably large numbers of mentally ill travellers who turned to begging, but had never been near Bedlam. It was adopted as a technique of begging, or a character. For example, Edgar in King Lear disguises himself as mad "Tom O'Bedlam". It was a popular enough ballad that another poem was written in reply, "Mad Maudlin's Search" or "Mad Maudlin's Search for Her Tom of Bedlam"[3] (the same Maud who was mentioned in the verse "With a thought I took for Maudlin / And a cruise of cockle pottage / With a thing thus tall, Sky bless you all / I befell into this dotage." which apparently records Tom going mad, "dotage") or "Bedlam Boys" (from the chorus, "Still I sing bonny boys, bonny mad boys / Bedlam boys are bonny / For they all go bare and they live by the air / And they want no drink or money."), whose first stanza was:

For to see Mad Tom of Bedlam,

Ten thousand miles I've traveled.

Mad Maudlin goes on dirty toes,

For to save her shoes from gravel

I went down to Satan's kitchen

To break my fast one morning

And there I got souls piping hot

All on the spit a-turning.

There I took a cauldron

Where boiled ten thousand harlots

Though full of flame I drank the same

To the health of all such varlets.

My staff has murdered giants

My bag a long knife carries

To cut mince pies from children's thighs

For which to feed the fairies.

No gypsy, slut or doxy

Shall win my mad Tom from me

I'll weep all night, with stars I'll fight

The fray shall well become me.

That into raggs would rend ye,

And the spirit that stands by the naked man

In the Book of Moones - defend ye!

That of your five sound senses

You never be forsaken,

Nor wander from your selves with Tom

Abroad to beg your bacon.

(Chorus; sung after every verse)

While I doe sing "any foode, any feeding,

Feedinge, drinke or clothing,"

Come dame or maid, be not afraid,

Poor Tom will injure nothing.

Of thirty bare years have I

Twice twenty been enraged,

And of forty been three times fifteen

In durance soundly caged.

On the lordly lofts of Bedlam,

With stubble soft and dainty,

Brave bracelets strong, sweet whips ding-dong,

With wholesome hunger plenty.

With a thought I took for Maudlin

And a cruse of cockle pottage,

With a thing thus tall, skie blesse you all,

I befell into this dotage.

I slept not since the Conquest,

Till then I never waked,

Till the roguish boy of love where I lay

Me found and stript me naked.

When I short have shorne my sowre face

And swigged my horny barrel,

In an oaken inn I pound my skin

As a suit of gilt apparel.

The moon's my constant Mistrisse,

And the lowly owl my morrowe,

The flaming Drake and the Nightcrow make

Me music to my sorrow.

The palsie plagues my pulses

When I prigg your pigs or pullen,

Your culvers take, or matchless make

Your Chanticleers, or sullen.

When I want provant, with Humfrie

I sup, and when benighted,

I repose in Powles with waking souls

Yet never am affrighted.

I know more than Apollo,

For oft, when he lies sleeping

I see the stars at bloody wars

In the wounded welkin weeping,

The moone embrace her shepherd

And the queen of Love her warrior,

While the first doth horne the star of morne,

And the next the heavenly Farrier.

The Gipsie Snap and Pedro

Are none of Tom's companions.

The punk I skorne and the cut purse sworne

And the roaring boyes bravadoe.

The meek, the white, the gentle,

Me handle touch and spare not

But those that crosse Tom Rynosseros

Do what the panther dare not.

With a host of furious fancies

Whereof I am commander,

With a burning spear and a horse of air,

To the wilderness I wander.

By a knight of ghostes and shadowes

I summon'd am to tourney

Ten leagues beyond the wild world's end.

Methinks it is no journey.

The existence of a chorus suggests that it was originally sung as a ballad. Both "Tom O' Bedlam" and "Mad Maudlin" are difficult to give a definitive form, because of the number of variant versions and the confusion between the two within the manuscripts.his version is taken from Bloom,

The religious priory of St Mary of Bethlem, in London, was confiscated by King Edward 3rd in 1375, and used for lunatics from 1377. In 1403/1404 it had just six insane patients and three who were sane. This old Bedlam was a small institution (on a site south of what is now Liverpool Street Station), even in the

17th century when it had about 30 patients. Its showy replacement, the Moorfields Bedlam, opened in 1676. A London hospital originally intended for the poor suffering from any ailment and for such as might have no other lodging, hence its name, *Bethlehem,* in Hebrew, the "house of bread." During the fourteenth century it began to be used partly as an asylum for the insane, for there is a report of a Royal Commission, in 1405, as to the state of lunatics confined there. The word *Bethlehem* became shortened to *Bedlam* in popular speech, and the confinement of lunatics there gave rise to the use of this word to mean a house of confusion. Bedlam was founded in 1247 as a priory in Bishopsgate Street, for the order of St. Mary of Bethlehem, by Simon Fitz Mary, an Alderman and Sheriff of London. This site is now occupied by the Liverpool Street railway station. In the next century it is mentioned as a hospital in a license granted (1330) to collect alms in England, Ireland, and Wales. In 1375 Bedlam became a royal hospital, taken by the crown on the pretext that it was an alien priory. It seems afterwards to have reverted to the city. At the beginning of the sixteenth century the word Bedlam was used by Tyndale to mean a madman, so that it would seem as though the hospital were now used as a lunatic asylum exclusively. In January, 1547, King Henry VIII formally granted St. Bartholomew's hospital and Bedlam, or Bethlehem, to the city of London, on condition that the city spends a certain amount on new buildings in connection with St. Bartholomew's. In 1674, the old premises having become untenable, it was decided to build another hospital, and this was erected in what is now Finsbury Circus. This came to be known as old Bedlam, after the erection of a new building in St. George's Fields, which was opened August 1815, on the site of the notorious tavern called the Dog and the Duck. The attitude of successive generations of Englishmen towards the insane can be traced interestingly at Bedlam. Originally, it was founded and kept by religious. Every effort seems to have been made to bring patients to such a state of mental health as would enable them to leave the asylum. An old English word, "a Bedlam" signifies one discharged and licensed to beg. Such persons wore a tin plate on their arm as a badge and were known as Bedlamers, Bedlamites, or Bedlam Beggars. Whenever outside inspection was not regularly maintained, abuses into the management of Bedlam, and in every century there were several commissions of investigation. Evelyn in his Diary, 21 April 1656, notes that he saw several poor creatures in Bedlam in chains. In the next century it became the custom for the idle classes to visit Bedlam and observe the antics of the insane patients as a novel form of amusement. This was done even by the nobility and their friends. One penny was charged for admission into the hospital, and there is a tradition that an annual income of four hundred pounds was thus realized. This would mean that nearly 100,000 persons visited the hospital in the course of a year. Hogarth's famous picture represents two fashionable ladies visiting the hospital as a show place, while his "Rake," at the end of the "Progress," is being fettered by a keeper. After an investigation in 1851, the hospital came under regular government inspection and has since been noted for its model care of the insane. It accommodates about three hundred, with over sixty attendants. Its convalescent home at Witley is an important feature. The management is so good that each year more than one-half of the patients are returned as cured.

Later periods

Church of the Holy Sepulchre (1885). Other than some restoration work, its appearance has essentially not changed since 1854. The same small ladder below the top-right window is also visible in recent photographs; this has remained in the same position since 1854 over a disagreement to move it. The Franciscan friars renovated it further in 1555, as it had been neglected despite increased numbers of pilgrims. The Franciscans rebuilt the Aedicule, extending the structure to create an ante-chamber. After the renovation of 1555, control of the church oscillated between the Franciscans and the Orthodox, depending on which community could obtain a favorable firman from the Sublime Porte at a particular time, often through outright bribery, and violent clashes were not uncommon. There was no agreement about this question, although it was talked about it at the negotiations to the Treaty of Karlowitz in 1699. In 1767, weary of the squabbling, the Porte issued a firman that divided the church among the claimants. A fire severely damaged the structure again in 1808, causing the dome of the Rotunda to collapse and smashing the Edicule's exterior decoration. The Rotunda and the Edicule's exterior were rebuilt in 1809–1810 by architect Komminos of Mytilene in the then current Ottoman Baroque style. The fire did not reach the interior of the Aedicule, and the marble decoration of the Tomb dates mainly to the 1555 restoration, although the interior of the ante-chamber, now known as the Chapel of the Angel, was partly re-built to a square ground-plan, in place of the previously semi-circular western end. Another decree in 1853 from the sultan solidified the existing territorial division among the communities and set a status quo for arrangements to "remain forever", caused differences of opinion about upkeep and even minor changes, including disagreement on the removal of an exterior ladder under one of the windows; this ladder has remained in the same position since then.

The church, after its 1808 restoration

The cladding of red marble applied to the Aedicule by Komminos has deteriorated badly and is detaching from the underlying structure; since 1947 it has been held in place with an exterior scaffolding of iron girders installed by the British Mandate. No plans have been agreed upon for its renovation. The current dome dates from 1870, although it was restored during 1994–1997, as part of extensive modern renovations to the church, which have been ongoing since 1959. During the 1970–1978 restoration works and excavations inside the building, and under the nearby Muristan, it was found that the area was originally a quarry, from which white meleke limestone was struck. To the east of the Chapel of Saint Helena, the excavators discovered a void containing a 2nd century drawing of a Roman ship, two low walls which supported the platform of Hadrian's 2nd century temple, and a higher 4th century wall built to support Constantine's basilica. After the excavations of the early 1970s, the Armenian authorities converted this archaeological space into the Chapel of

Saint Vartan, and created an artificial walkway over the quarry on the north of the chapel, so that the new Chapel could be accessed (by permission) from the Chapel of Saint Helena. There was some controversy in 2010, when the Jerusalem City Council threatened to cut off water to the site, due to disputed water bills. The entrance to the church is through a single door in the south transept. This narrow way of access to such a large structure has proven to be hazardous at times. For example, when a fire broke out in 1840, dozens of pilgrims were trampled to death. In 1999 the communities agreed to install a new exit door in the church, but there was never any report of this door being completed.

The Altar of the Crucifixion

On the south side of the altar via the ambulatory (an aisle surrounding the end of the choir or chancel of a church) is a stairway climbing to Calvary (Golgotha), traditionally regarded as the site of Jesus' crucifixion and the most lavishly decorated part of the church.The main altar there belongs to the Greek Orthodox, which contains The Rock of Calvary (12th Station of the Cross). The rock can be seen under glass on both sides of the altar, and beneath the altar there is a hole said to be the place where the cross was raised. Due to the significance of this, it is the most visited site in the Holy Sepulchre. The Roman Catholics (Franciscans) have an altar to the side, The Chapel of the Nailing of the Cross (11th Station of the Cross). On the left of the altar, towards the Eastern Orthodox chapel, there is a statue of Mary, believed to be working wonders (the 13th Station of the Cross, where Jesus' body was removed from the cross and given to his family). Beneath the Calvary and the two chapels there, on the main floor, there is The Chapel of Adam. According to tradition, Jesus was crucified over the place where Adam's skull was buried. The Rock of Calvary is seen cracked through a window on the altar wall, the crack traditionally being said to be caused by the earthquake that occurred when Jesus died on the cross, and being said by more critical scholars to be the result of quarrying against a natural flaw in the rock.

The Rock of Calvary seen under glass

Just inside the entrance is The Stone of Anointing, also known as The Stone of Unction, which tradition claims to be the spot where Jesus' body was prepared for burial by Joseph of Arimathea. However, this tradition is only attested since the crusader era, and the present stone was only added in the 1810 reconstruction. The wall behind the stone was a temporary addition to support the arch above it, which had been weakened after the damage in the 1808 fire; the wall blocks the view of the rotunda, sits on top of the graves of four 12th century kings, and is no longer structurally necessary. There is a difference of opinion as to whether it is the 13th Station of the Cross, which others identify as the lowering of Jesus from the cross and locate between the 11th and 12th station up on Calvary. The lamps that hang over the stone are contributed by Armenians, Copts, Greeks and Latins.

The Rotunda and the Aedicule

The Rotunda is located in the centre of the Anastasis, beneath the larger of the church's two domes. In the centre of the Rotunda is the chapel called The Edicule, which contains the Holy Sepulchre itself. The Edicule has two rooms. The first one holds The Angel's Stone, a fragment of the stone believed to have sealed the tomb after Jesus' burial. The second one is the tomb itself.

The Status Quo in the Rotunda

Under the status quo, the Eastern Orthodox, Roman Catholic, and Armenian Apostolic Churches all have rights to the interior of the tomb, and all three communities celebrate the Divine Liturgy or Holy Mass there daily. It is also used for other ceremonies on special occasions, such as the Holy Saturday ceremony of the Holy Fire celebrated by the Greek Orthodox Patriarch of Jerusalem. To its rear, within a chapel constructed of iron latticework upon a stone base semicircular in plan, lies the altar used by the Coptic Orthodox. Beyond that to the rear of the Rotunda is a very rough hewn chapel, containing an opening to a rock-cut chamber, from which several kokh-tombs radiate.

Although this space was discovered comparatively recently, and contains no identifying marks, many Christians believe it to be the tomb of Joseph of Arimathea in which the Syriac Orthodox celebrate their Liturgy on Sundays. To the right of the sepulchre on the southeastern side of the Rotunda is the Chapel of the Apparition which is reserved for Roman Catholic use.

The omphalos and the north wall of the Catholicon

The Catholicon On the east side opposite the Rotunda is the Crusader structure housing the main altar of the Church, today the Greek Orthodox catholicon. The second, smaller dome sits directly over the centre of the transept crossing of the choir where the compas, an omphalos once thought to be the centre of the world (associated to the site of the Crucifixion and the Resurrection), is situated. East of this is a large iconostasis demarcating the Orthodox sanctuary before which is set the throne of the Greek Orthodox Patriarch of Jerusalem on the south side facing the throne of the Greek Orthodox Patriarch of Antioch on the north side.

The "Holy Prison", or Prison of Christ

Prison of Christ - In the north-east side of the complex there is The Prison of Christ, alleged by the Franciscans to be where Jesus was held. The Greek Orthodox allege that the real place that Jesus was held was the similarly named Prison of Christ, within their Monastery of the Praetorium, located near the Church of Ecce Homo, at the first station on the Via Dolorosa. The Armenians regard a recess in the Monastery of the Flagellation, a building near the second station on the Via Dolorosa, as the Prison of Christ. A cistern among the ruins near the Church of St. Peter in Gallicantu is also alleged to have been the Prison of Christ. Further to the east in the ambulatory are three chapels (from south to north): Greek Chapel of St. Longinus - The Orthodox Greek chapel is dedicated to St. Longinus, a Roman soldier who according the New Testament pierced Jesus with a spear.

Chapel of Saint Helena, Jerusalem

The Chapel of Saint Helena - between the first two chapels are stairs descending to The Chapel of Saint Helena, belonging to the Armenians Chapel of Saint Vartan - on the north side of the Chapel of Saint Helena is an ornate wrought iron door, beyond which a raised artificial platform affords views of the Quarry, and which leads to the Chapel of Saint Vartan. The latter Chapel contains archaeological remains from Hadrian's temple and Constantine's basilica. These areas are usually closed. Chapel of the Invention of the Holy Cross - another set of 22 stairs from the Chapel of Saint Helena leads down to the Roman Catholic Chapel of the Invention of the Holy Cross believed to be the place where the True Cross was found.

North of the Aedicule

The Franciscan Chapel of St. Mary Magdalene - The chapel indicates the place where Mary Magdalene met Jesus after his resurrection. The Franciscan Chapel of the Blessed Sacrament (or Chapel of the Apparition) in memory of Jesus' meeting with his mother after the Resurrection

South of the Aedicule

The three Greek Orthodox chapels of St. James the Just, St. John the Baptist and of the Forty Martyrs of Sebaste, south of the rotunda and on the west side of the front courtyard originally formed the baptistery complex of the Constantinean church, the southern most chapel being the vestibule, the middle chapel being the actual baptistery and the north chapel being the chamber in which the patriarch chrismated the newly baptized before leading them into the rotunda north of this complex. The primary custodians are the Eastern Orthodox, Armenian Apostolic, and Roman Catholic Churches, with the Greek Orthodox Church having the lion's share. In the 19th century, the Coptic Orthodox, the Ethiopian Orthodox and the Syriac Orthodox acquired lesser responsibilities, which include shrines and other structures within and around the building. Times and places of worship for each community are strictly regulated in common areas. Establishment of the 1853 status quo did not halt the violence, which continues to break out every so often even in modern times. On a hot summer day in 2002, a Coptic monk moved his chair from its agreed spot into the shade. This was interpreted as a hostile move by the Ethiopians, and eleven were hospitalized after the resulting fracas. In another incident in 2004, during Orthodox celebrations of the Exaltation of the Holy Cross, a door to the Franciscan chapel was left open. This was taken as a sign of disrespect by the Orthodox and a fistfight broke out. Some people were arrested, but no one was seriously injured.

Franciscans during the procession on The Calvary, 2006

On Palm Sunday, in April 2008, a brawl broke out when a Greek monk was ejected from the building by a rival faction. Police were called to the scene but were also attacked by the enraged brawlers.[31] On Sunday, November 9, 2008, a clash erupted between Armenian and Greek monks during celebrations for the Feast of the Cross. Under the status quo, no part of what is designated as common territory may be so much as rearranged without consent from all communities. This often leads to the neglect of badly needed repairs when the communities cannot come to an agreement among themselves about the final shape of a project. Just such a disagreement has delayed the renovation of the edicule, where the need is now dire, but also where any change in the structure might result in a change to the status quo, disagreeable to one or more of the communities. A less grave sign of this state of affairs is located on a window ledge over the church's entrance. Someone placed a wooden ladder there sometime before 1852, when the status quo defined both the doors and the window ledges as common ground. The ladder remains there to this day, in almost exactly the same position it can be seen to occupy in century-old photographs and engravings. An engraving by David Roberts in 1839 also shows the same ladder in the same position. No one controls the main entrance. In 1192, Saladin assigned responsibility for it to the Muslim Nuseibeh family. The Joudeh Al-Goudia family were entrusted as custodian to the keys of the Holy Sepulchre by the Ottomans few hundred years later, and both families now share the responsibility. This arrangement has persisted into modern times.

Relationship to Temple of Aphrodite

Jerusalem after being rebuilt by Hadrian. Two main east-west roads were built rather than the typical one, due to the awkward location of the Temple Mount, blocking the central east-west route. The site of the Church had been a temple of Aphrodite prior to Constantine's aedifice being built. Hadrian's temple had actually been located there because it was the junction of the main north-south road with one of the two main east-west roads and directly adjacent to the forum (which is now the location of the (smaller) Muristan); the forum itself had been placed, as is traditional in Roman towns, at the junction of the main north-south road with the (other) main east-west road (which is now El-Bazar/David Street). The temple and forum together took up the entire space between the two main east-west roads (a few above-ground remains of the east end of the temple precinct still survive in the Russian Mission in Exile). From the archaeological excavations in the 1970s, it is clear that construction took over most of the site of the earlier temple enclosure and that the Triportico and Rotunda roughly overlapped with the temple building itself; the excavations indicate that the temple extended at least as far back as the Aedicule, and the temple enclosure would have reached back slightly further. Virgilio Canio Corbo, a Franciscan priest and archaeologist, who was present at the excavations, estimated from the archaeological evidence that the western retaining wall, of the temple itself, would have passed extremely close to the east side of the supposed tomb; if the wall had been any further west any tomb would have been crushed under the weight of the wall (which would be immediately above it) if it had not already been destroyed when foundations for the wall were made. Other archaeologists have criticised Corbo's reconstructions. Dan Bahat, the former city archaeologist of Jerusalem, regards them as unsatisfactory, as there is no known Temple of Aphrodite matching Corbo's design, and no archaeological evidence for Corbo's suggestion that the

Temple Building was on a platform raised high enough to avoid including anything sited where the Aedicule is now; indeed Bahat notes that many temples to Aphrodite have a rotunda-like design, and argues that there is no archaeological reason to assume that the present rotunda wasn't based on a rotunda in the temple previously on the site.

Relationship to city walls

The Bible describes Jesus' tomb as being outside the city wall, as was normal for burials across the ancient world, which were regarded as unclean, but the Church of the Holy Sepulchre is in the heart of Hadrian's city, well within the Old City walls, which were built by Sultan Suleiman the Magnificent in 1538. Some have claimed that the city had been much narrower in Jesus' time, with the site then having been outside the walls; since Herod Agrippa is recorded by history as extending the city to the north (beyond the present northern walls), the required repositioning of the western wall is traditionally attributed to him as well. If the western city wall was originally to the east of the Church of the Holy Sepulchre, then the western hill, on which it is sited, would have been advantageous to an enemy. However, a wall would imply the existence of a defensive ditch outside it, so an earlier wall could not be immediately adjacent to site of the tomb, which combined with the presence of the Temple Mount would make the city inside the wall quite thin; essentially for the traditional site to have been outside the wall, the city would have had to be limited to the lower parts of the Tyropoeon Valley, rather than including the defensively advantageous western hill. Since these geographic considerations imply that, not including the hill within, the walls would be willfully making the city prone to attack from it, some scholars, including the late 19th century surveyors of the Palestine Exploration Fund, consider it unlikely that a wall would ever have been built that would cut the hill off from the city in the valley; archaeological evidence for the existence of an earlier city wall in such a location has never been found. The area immediately to the south and east of the sepulchre was a quarry[] and outside the city during the early 1st century as excavations under the Lutheran Church of the Redeemer across the street demonstrated. This obviated the need for a defensive ditch or fosse since the line of the city wall would follow the south lip of the quarry. The quarry and tombs associated with it are north, not west of the main city and west only of the merchant area in the Tyropoeon Valley, which was enclosed by the Second Wall. Although, in 2007, Dan Bahat stated that Six graves from the first century were found on the area of the Church of the Holy Sepulchre. That means, this place was outside of the city, without any doubt, the dating of the tombs is based on the fact that they are in the kokh style, which was common in 1st century; however, the kokh style of tomb was also common in the first to 3rd centuries BC. The likelihood of a 1st century tomb being built to the west of the city is questionable, as according to the late 1st century Rabbinic leader, Akiva ben Joseph, quoted in the Mishnah, tombs should not built to the west of the city, as the wind in Jerusalem generally blows from the west, and would blow the smell of the corpses and their impurity over the city, and the Temple Mount. Additionally, the Aedicule would be quite close to the city even the west wall of the city had been to its east; yet Akiba remarks that Jewish law insists that tombs should not be built within 50 cubits of a city. The archaeological record indicates that the instructions reported by Akiba, for choosing a burial location, were rigidly adhered to; almost all of the tombs from classical Jerusalem are to the east of the city, on the Mount of Olives, except for a few located over a kilometre to the west, and those in the Church of the Holy Sepulchre. One might note, however, that what is assumed to be a niche for the Torah scroll in the building probably originally built as a Judeo-Christian synagogue between AD 70 and AD 135 on the traditional site of the Cenacle or upper room of the Last Supper and now identified as the site of the King David's Tomb is oriented not towards the Temple Mount, but towards the site of the Holy Sepulchre, which would seem to indicate that the Christian community that had built it had already began to transfer many of the religious traditions originally associated with the Temple to the sites they associated with Christ's death and resurrection (such as the burial place of Adam and the centre of the world)

Challenges to authenticity

Although the identification of the Aedicule as the site of Jesus' tomb is not a tenet of faith for any major Christian denomination, many Catholic and Orthodox Christians hold fast to this traditional location. However, due to the many issues the site raises, several scholars have rejected its validity. Additionally many Protestants have often opposed the traditional location because it has previously received support from Catholics. After time spent in Palestine in 1882–83, General Charles George Gordon found a location outside the old city walls that he suggested to have been the real location of Golgotha. Although the Church of the Holy Sepulchre has its tomb just a few yards away from its Golgotha, there is no particular reason to regard this close juxtaposition as a necessity; however, Gordon followed this principle, concluding that his site for Golgotha must also be the approximate location for Jesus' burial, identifying a nearby tomb, now called the Garden Tomb, as the location for the event. Pottery and archaeological findings in the area have subsequently been dated to the 7th century BC so, in the opinion of archaeologists the Garden Tomb site would have been abandoned by the 1st century. Biblically this does not match three of the Gospel accounts (Matthew, Luke, and John) which specifically state the tomb was new and no one had ever been laid inside. Despite the archaeological discoveries, the Garden Tomb has become a popular place of pilgrimage among Protestants. The Church of Jesus Christ of Latter-day Saints leaders have been more hesitant to formally commit to the identification even though many Mormons regard the Garden Tomb as the correct location of Jesus' tomb. From the 9th century, the construction of churches inspired in the Anastasis was extended across Europe. One example is Santo Stefano in Bologna, Italy, an agglomeration of seven churches recreating shrines of Jerusalem. Several churches and monasteries in Europe, for instance, in Germany and Russia, have been modeled on the Church of the Resurrection, some even reproducing other holy places for the benefit of pilgrims who could not travel to the Holy Land. They include the Heiliges Grab of Görlitz, constructed between 1481 and 1504, and the New Jerusalem Monastery in Moscow Oblast, constructed by Patriarch Nikon between 1656 to 1666.

St. Bethlehem Hosital the world's first Mental Hospital

London's St Bethlehem Hospital was the world's first institution dedicated to mental illness. An abbreviation of its name produced the modern word "bedlam". The institution was perhaps never very large. It was certainly of much less importance than the other house outside Bishopsgate. The respective spheres of St. Mary Spital and the rector of St. Botolph's had had to be determined within a few years of the foundation of the priory, but it was not until 1362 that the building of a chapel in honour of the Virgin and the Nativity of Jesus by the house of St. Mary of Bethlehem made an agreement between the rector and this hospital necessary. By the arrangement then made the master and brethren were permitted to complete the chapel, have bells rung there, celebrate divine service, and receive offerings; they might also bury any who wished to be buried in the chapel or precincts, and have the oblations or obviations, except in the case of parishioners of St. Botolph's, when half the offering was to go to the rector. Considering that at this time their fixed income was only 33s. per annum, and that the proceeds of the collection, which by royal license they made throughout the kingdom, had probably fallen off after the plague of 1350, this settlement was important, and in order to swell the flow of offerings they obtained from the pope in 1363 a special indulgence, extending over a period of ten years, to those who at Christmas, the Epiphany, and the five feasts of the Virgin Mary, with their vigils, visited and rendered material aid to the hospital. In 1389 it benefited, presumably to the extent of £100, by the will of Ralph Basset of Drayton who erected two chantries there. It must also have reaped some advantage from a gild called the Fraternity of St. Mary of Bethlehem established in the church in 1370. The connexion of the house with the bishopric of Bethlehem doubtless came to an end in the latter part of the thirteenth century, when the Holy Land was lost to Christendom, but how or when the king obtained the patronage it is impossible to say. The corporation of London in 1346 took the hospital under its protection, and had certainly some kind of right over the place in 1350, for on the death of the master, John de Nortone, the sergeant was ordered to take possession of the house in the name of the City, though the order was afterwards rescinded because the hospital had been let to a certain Robert Aaunsard, fishmonger, for a term of years. In 1381, when the king appointed William Welles as master, the City disputed his right, asserting that the hospital was in their gift. At first they were successful, but in the end the crown gained the day, and appointed as in the case of a royal free chapel, which the hospital resembled also in another point, viz. its exemption from the jurisdiction of the ordinary. Some very interesting facts about the house were disclosed during a visitation by two of the king's clerks in March, 1403. It had already become an asylum principally, though not exclusively, for the insane, and at that time there were six lunatics and three sick persons there. These people, or their relatives, contributed something to their support, but the amount varied, the highest rate mentioned being 12d. A week paid by a merchant of Exeter, who was there for six weeks. The hospital had a little property, but was chiefly maintained by voluntary contributions, and it was calculated that the collections throughout England brought in about 40 marks a year, the obviations in the great and small chapels 52s., those on the great feasts another 52s., the box at the door of the house and the two boxes carried about London and the suburbs similar amounts, and the offerings for the poor on the day of the Parascene 20s. A collection throughout the diocese of London for the sick poor amounted roughly to 4 marks annually, and gifts of meat, ale, fish, salt, and candles were also made. The management of the hospital appears at this time to have belonged to the office of porter, and Peter Taverner, who had received the post for life, had abused his trust in every way. He had rendered no accounts of the money accruing from the various collections, in some cases for four years, in others for fourteen, nor of bequests and payments made for the inmates. He had not distributed the alms, but with the money had bought fuel and made the poor pay for it, while his wife had taken the best of the contributions in kind. Not content with this, he had disposed of the beds and other goods, causing a loss to the hospital of about £40, and through him robbers had caused even worse damage. In spite of the remonstrances of the master he persisted in playing at dice and draughts, and in selling ale at his house within the close. It is incredible that Taverner's conduct would have been so long unchecked if the master had been constantly resident or really interested in the place, and it may be noted that the statement of one of the inmates that divine service was sometimes withdrawn by the default of the master or his curate was found to be true, and that the chapel was but poorly provided with books and plate, while it was also said to be his fault that there were no brothers and sisters in the hospital. The distinctive dress of the order had been abandoned, and with it seems to have vanished most of the character of the original foundation. Some kind of reconstitution must have been affected, since in 1424 brethren and sisters were associated with the master in sending a proctor or quaestor to seek alms in the archdeaconry of Oxford. But it is evident that in one important respect the hospital developed in the direction it had already taken in. The fourteenth century, the office of master tending more and more to become a sinecure. Proof of this may probably be found in the hospital being let to farm by its head in 1454, but there can be no doubt of the significance of the appointment of George Boleyn, a layman, in 1529, and on his forfeiture of a gentleman of the privy chamber. In 1523 Stephen Gennings, a merchant-tailor gave £40 to the City Corporation towards the purchase of the patronage of the house, which, however, was not effected until 1546. As there is no Valor there are no means of ascertaining what property the hospital had at this date, but the income derived from it seems to have been less than £40, and was so inadequate to the demands upon it that recourse was had in 1551 to the old practice of soliciting alms of the charitable, in this instance within the counties of Lincoln and Cambridge, the isle of Ely, and the city of London. In 1632 commissioners were appointed to inquire into the state of the hospital, which was found to be very unsatisfactory. A sum of 2s. a week was allowed for each patient, but as the master, Dr. Crooke, spent most of it on himself, and the steward appropriated the gifts in kind, the unfortunate inmates, unless they bought of the steward at extortionate rates, were almost starved. It need hardly be added that no measures were taken to cure them of their malady. The income of the house was £277 3s. 4d. but this did not include the weekly donations of food from the Lord Mayor, sheriffs, and other persons. There appears to have been some idea of enlarging the hospital in 1644, a project made impossible by the Civil War, which diminished its revenues and caused it to be converted to other uses for the time being. In 1675, however, the increased number for whom admission was requested made larger quarters a necessity, and as the situation of the house was not a good one for the purpose, a new hospital was built in 1675, at a cost of nearly £17,000, on ground in Moorfields granted by the City. At that time it formed one corporation with Bridewell, and the superior officials were common to the two institutions, but each had a committee of its own, and a subdivision of this went to the hospital once a week to check the accounts and inspect the food. The building was much enlarged in 1734, when accommodation was provided for 100 incurable cases as well as for more patients not supposed to be hopeless. The inspecting committee evidently worked well, and the management of the place was excellent. Care was taken to make the charges on friends of the patients as small as possible, and the welfare of the lunatics was the chief consideration. Early sixteenth century maps show Bedlam, next to Bishopsgate, as a courtyard with a few stone buildings, a church and a garden. Conditions were consistently dreadful, and the care amounted to little more than restraint. There were 31 patients and the noise was "so hideous, so great; that they are more able to drive a man that hath his wits rather out of them." Violent or dangerous patients were manacled and chained to the floor or wall. Some were allowed to leave, and licensed to beg. It was a Royal hospital, but controlled by the City of London after 1557, and managed by the Governors of Bridewell. Day to day management was in the hands of a Keeper, who received payment for each patient from their parish, livery company, or relatives. In 1598 an inspection showed neglect; the "Great Vault" (cesspit) badly needed emptying, and the kitchen drains needed replacing. There were 20 patients there, one of whom had been there over 25 years. Many sources assert that, in 1620, patients of Bethlem banded together and sent a "Petition of the Poor Distracted People in the House of Bedlam (concerned with conditions for inmates)" to the House of Lords. However, the absence of a primary document suggests that this may well be a 'phantom reference'. The Hospital became famous and notorious for the brutal ill-treatment meted out to the mentally ill. In 1675 Bedlam moved to new buildings in Moorfields designed by Robert Hooke, outside the City boundary. The playwright Nathaniel Lee was incarcerated there for five years, reporting that: "They called me mad, and I called them mad, and damn them, they outvoted me." The inmates were first called "patients" in 1700, and "curable" and "incurable" wards were opened in 1725-34. In the 18th century people used to go to Bedlam to stare at the lunatics. For a penny one could peer into their cells, view the freaks of the "show of Bethlehem" and laugh at their antics. Entry was free on the first Tuesday of the month. In 1814 alone, there were 96,000 such visits.

James Monro (1680-1752): the First in Line

(Physician to Bethlem Royal Hospital 1728-1752)

From 1632 Bethlem, in line with the practice already employed at other royal hospitals, appointed three medical officers elected by the Court of Governors physician, surgeon and apothecary. Competition for the post of physician was fierce and required an Oxbridge education and affiliation with the College of Physicians - but was often determined by influence rather than qualifications. James Monro, although well known for his pioneering experiments in smallpox inoculation, was not the obvious choice in 1728. In the medical profession at this time it was not abnormal for a son to inherit his father's practice. The reign of over 120 years by the Monro dynasty as physicians at Bethlem was, however, unprecedented. Yet each successive Monro did have to prove his worth to the Governors prior to being appointed. Bethlem physicians were initially appointed merely in a visiting capacity and paid a nominal fee. Riches came from private practice. The Monros were also owners of private 'madhouses' in Hackney and Clerkenwell, and visiting physicians to scores of others. Indeed the family name 'Monro', became synonymous with the very word 'mad-doctor'. Absenteeism of the medical officers at Bethlem led to patients too often being at the mercy of untrained and overstretched staff. This was a central point of the shocking revelations of the Parliamentary enquiry into 'madhouses' in 1815/16. James Monro, and his son John, had already suffered a public scandal with the publication of 'A Treatise on Madness' in 1758. Here, William Battie, physician to the London asylum of St. Luke's, argued that Bethlem and the treatment doled out by the Monros, was antiquated, negligent and abusive. So began the great rivalry between the two establishments.

John Monro, (1715-1791)

Son of James Monro: the Golden Age of 'Mad-Doctoring'

(Physician to Bethlem Royal Hospital 1752-1791)

With the series of mental disorders experienced by King George III, which began in 1789, 'mad-doctoring' was for the first time publicly highlighted as a necessary, if not wholly respectable, medical specialist. At the time, John Monroe was the most prominent medical man in the newly emerging profession of managing 'the mad'. He succeeded his father as physician to the notorious Bethlem Hospital, was consulted over the King's mental malady, and enjoyed high office at the Royal College of Physicians - presenting the Hardeman Oration in 1757, and acting as Censor seven times. It was John Monro's private practice that consumed the majority of his time a practice that developed to fulfill the growing demands of the middle and upper classes for discreet private care of the insane. As consultant to patients in a number of London's private 'mad-houses', and proprietor of Brooke House in Hackney, Monro enjoyed a lucrative and thriving business. One very rare, and possibly unique, document chronicling the private medical care of the mentally ill of the 18th century survives John Monro's case book of 1766. Monro did, however, appear more dedicated than his father, James, to his role as physician at Bethlem. He advised on Committees that were central in instigating changes such as adding new cells for patients and curtailing public visiting in 1769/70. However, his response to the public criticism of Bethlem's treatments as violent and shocking, by William Battie, physician to St. Luke's Asylum, was both negative and complacent. With this establishment opening its doors in 1751, Bethlem lost its monopoly as the sole establishment for treating the insane in London. John Monro, rather than seeing it as an opportunity for collaboration and partnership, dug his heels in and defended his family's traditional evacuative therapies – those of purges and vomits. In reality this treatment was still in use at St. Luke's and was following the approved medical practices of the day.

Thomas Monro (1759-1833)

(Physician to Bethlem Royal Hospital 1791-1816)

During the 18th century there was a gradual shift in public attitude towards the treatment of the insane, and a more compassionate view emerged. The controversial treatment applied to King George III during his bouts of madness, was well publicized. Even the King was not above shock therapies. The publication in 1813 of Samuel Tuke's 'Description of the Retreat' in York, with its non-restraint and moral management of kindness and discipline, further swayed public opinion. The Retreat was held up as a paradigm of virtue and pressure for reform mounted. The lunacy reform movement took up the gauntlet and two institutions fell under attack: York Asylum and Bethlem Royal Hospital. A complaint of ill treatment of a patient, William Vickers, at York Asylum in 1813, led to a full examination. Scandals of abuse and embezzlement of funds were unearthed and enthusiastically transmitted to the public by the York Herald. Bethlem was next. In 1814, Edward Wakefield, a leader of the lunacy reform movement, began investigating, a year before the hospital moved into its new premises at St. George's Field, Southwark. The dilapidated Moorfields building was not Bethlem at its best. Yet it was the treatment of patients, not the physical building that was the most damning element of Wakefield's report. The discovery of James Norris, restrained by chains for 14 years within Bethlem, epitomized for Wakefield all that was wrong with the institution: barbaric, inhumane and outdated practices. The evidence collected from Bethlem and York Asylum forced a Parliamentary investigation into 'madhouses' and the appointment of the 1815 Select Committee. Under cross examination by the Committee, Bethlem's physician Thomas Monro, and the apothecary John Haslam, incriminated themselves and each other, and became the sacrificial victims of the enquiry. But also exposed was a disturbing fact: this scandalous treatment of the 'insane' was in no way confined to Bethlem or York Asylum.

Edward Thomas Monro (1790-1856)

Son of Thomas Monro: the End of an Era

(Physician to Bethlem Royal Hospital 1816-1855)

The years following 1815, when Bethlem moved to new premises at St. George's Field, Southwark, and the Select Committee on Madhouses published its first report, were seen as a new enlightened chapter for the hospital. The evidence presented by the Committee demanded considerable change – in the medical treatment, building conditions, and management. Change took place. The disgraced physician Thomas Monro was re-appointed with none other than his son Edward Thomas Monro. A second physician, George Leman Tuthill - succeeded by Alexander Morison - was appointed to pacify those Governors who questioned the ongoing reign of the Monro dynasty. In an attempt to shake off the cloud of secrecy surrounding Bethlem, Monro and Tuthill were instructed to write an annual report. Their duties were more closely regulated - a physician was to attend the hospital on a daily basis. The shamed apothecary John Haslam was replaced by Edward Wright, and the role was elevated to one of superintendence. Under the physician ship of Alexander Morison, Bethlem saw more improvements. Morison was the first

Bethlem physician to embrace the concept of teaching 'insanity' to medical students. In following the progressive Hanwell Asylum's example, Bethlem, in 1842, became one of the first hospitals treating mental illness to admit students. Between 1815 and 1851, Bethlem was a tale of two halves: embracing contemporary treatment practiced at other asylums, whilst also continuing with traditional therapies. Bethlem had not witnessed its last scandal. Reports of patient neglect in 1851 led to Bethlem finally being included in the national system of investigation into asylums and hospitals for the mentally ill. Following the resignation in 1855 of Edward Thomas Monro, a period of reform followed which led to more enlightened treatment under new management. After more than 120 years, the final curtain had fallen on the reign of the Monros.

Bedlam Relics

Certain precious relics, which fell to it as a share of the spoils from the sack of Constantinople in 1204. These were the hammer and a nail of the crucifixion, with the hand of St. Thomas. Even these had gone into the sack with the title deeds of the convent lands, and were pledged with the Templars and other religious orders for a considerable sum of money. It is not pleaded on behalf of these sacrilegious rogues that they spent it on the ransom of the church and the pilgrim : they just " dissipated it themselves." We may note here that in 1869 some old relics of the church a pair of chandeliers and two copper basins of twelfth-century work were dug up in an ancient cloister which formed part of the Franciscan church of Bethlehem. The base of each chandelier was inscribed : " Cursed be he who removes me from the place of the Holy Nativity, Bethlehem." Possibly this is an allusion to the depre- dations of John the Roman, and they may have been returned to the basilica by Godfrey, who helped to found the priory of Bethlehem, London, for it was he who had to restore a church very much injured by " men who know not the way of the Lord," and to redeem as much of the treasure and property of the basilica as was still in the market.

James Tilly Matthews: A patient in Bethlem Royal Hospital 1797 to 1814

When the second home of Bethlem Royal Hospital opened at Moorefield's in 1676, it was praised as 'the only building that looked like a palace in London'. It was one of the great spectacles on any tour of the city. Yet by 1816, having become old and dilapidated, the building was pulled down. In 1810 an advertisement in 'The Times' offered prizes of £200, £100 and £50 for the best three designs for a new home for Bethlem. Of over thirty entries, one was of particular significance: that of James Tilly Matthews, a patient inside Bethlem. Well aware of the downfalls of Bethlem's old building, Matthews, a skilled draftsman, presented his own 46-page dossier of plans and explanations. Matthews was sent to Bethlem in 1797 by the Privy Council, following his disruption of the House of Commons with rantings of a French plot to mesmerize the British nation. Seen as a threat to King and country, Matthews remained at Bethlem in the incurable wing until 1814, the year before his death. John Haslam, Bethlem's apothecary, in providing proof of Matthews' insanity wrote 'Illustrations of Madness', published in 1810. Focusing on one individual, and including Matthews' own writings and drawings, this was the first work of its kind. The case of Matthews, and that of James Norris, became central to the scandalous Parliamentary enquiry of 1815/16. Whereas Norris, chained for 14 years within Bethlem, epitomized the brutal, Matthews' case revealed inequalities of class Matthews being 'a man of considerable accomplishments' and possessing 'great learning' was described as an 'unfit person for confinement in Bethlem'. Although Matthews' plans for the new Bethlem hospital were not used, he received £30 for his efforts. An act of genuine goodwill by the Governors a considerable sum at the beginning of the 19th century.

Michel Foucoult in his "Madness and Civilization" also the title has been translated "History of Madness" argues that in classical and Renaissance time people with mental problems were seen as possessing a kind of wisdom that is not from this world, and it was in the middle of the 17th century when problems like that started to be seen as unreasonable without making a distinction between psychological problems being embedded in the operation of the psyche, not-biological/ and biologically based ones. Hence, people with any kind of visible problems were separated from society and treated in the same way as common criminals.

The first Bethlem lunatics are recorded in 1403, possibly transferred from an existing site at Charing Cross. Bethlem then housed six deranged men: an inventory listed four pairs of manacles, eleven chains, six locks and two pairs of stocks (though these may not all have been for the lunatics). These links with lunacy, accidental at first, proved lasting. Early records note various donations for `the sick and insane', and in 1436 a tailor gained exemption from watch and jury service by pleading that he was required to attend `on the poor frenzied and demented creatures'. There was no obligation for the Master to be a medical man Knowledge of the patients is also bitty, and the darkness becomes visible only when something went wrong. In 1619 Dr Hilkiah Crooke, a Cambridge graduate and physician to James I, was appointed Master. Evidently ambitious, he urged that Bethlem should be made independent from Bridewell. But he proved negligent and grasping, and complaints were laid before the governors. In 1632 Charles I launched an investigation and, not for the last time, Bethlem won unwelcome exposure. Funds were flooding into the eponymous Crooke's pocket; accused of absenteeism, he countered that he had originally cured seventeen lunatics but had ceased because `the governors of Bridewell doe refuse to pay him his Apothecaries bills' -- a charge they denied. Crooke was dismissed; the scandal led to reorganization; record-keeping improved and, from then on, light dawns. Bethlem acquired physicians of some standing, including Edward Tyson, pioneer anatomist of the orangutan. Despite modern assumptions about malpractices, Bethlem's committee, comprising City worthies and meeting weekly, performed quite conscientiously. If all too often it had to reprimand the porter or steward or to sack the basket-men and gallery-maids, it was at least energetic in hounding endemic abuses.

Standards were set, and within the hospital orders were posted, including the following: V. that no Person will give the Lunatics Strong Drink. That such of the Lunatics as are fit, be permitted to walk in the Yard till Dinnertime. No Servant, or other Person whatsoever, shall take any Money given to the Lunatics. That some of the Committee goes weekly to the said Hospital, to see the Provision weighed. These rules were, of course, broken a pamphlet published in 1818 by an ex-patient, Urbane Metcalf, documented bullying and pilfering; but they contradict those historians who have claimed that Bethlem regarded the madmen as beasts, and accordingly the insane were chained and whipped. On the contrary, the committee forbade violence, insisting that patients were there for treatment not punishment. Unfortunately, this was what passed for treatment in those days: When Bethlem was visited in 1814 by the philanthropist, Edward Wakefield; he was shocked on encountering one patient. James Norris: A stout iron ring was riveted round his neck, from which a short chain passed through a ring made to slide upwards and downwards on an upright massive iron bar, more than six feet high, inserted into the wall. Round his body a strong iron bar about mo inches wide was riveted; on each side of the bar was a circular projection; which being fashioned to and enclosing each of his arms, pinioned them close to his sides. Norris had been thus immobilized for twelve years. When Wakefield publicized this outrage, a committee of the governors was convened, who came up with the hot air which institutions have perfected: Every attention has ... been paid in the Hospital to the cleanliness, the health, and the comfort of the patients confined therein, and ... every degree of indulgence consistent with the security of the patients and the safety of those employed has been observed. Evidence heard before the Commons committee revealed shocking, horrifying facts. `We first proceeded to visit the women's galleries', testified Wakefield: One of the side rooms contained about ten patients. Each chained by one arm or leg to the wall. The chain allowing them merely to stand up by the bench or form fixed to the wall, or to sit down on it. The nakedness of each patient was covered by a blanket-gown only. Monroe sought to reassure the gentlemanly MPs by stating that chains and fetters were `fit only for the pauper lunatics: if a gentleman was put in irons he would not like it'.

Eighteenth Century Bethlem

Eighteenth century Bethlem was most notably portrayed in a scene from William Hogarth's A Rake's Progress (1735), the story of a rich merchant's son whose immoral living causes him to end up in a ward at Bethlem. This reflects the view of the time that madness was a result of moral weakness, leading to "moral insanity" being used as a common diagnosis. In 1815, Bedlam was moved to St George's Fields, Southwark, into buildings designed by James Lewis (a cupola was added later by Sydney Smirke). The inmates were referred to as "unfortunates" and must have had an uncomfortable time in their first winter there; no glass was initially provided for the windows, because of "the disagreable effluvias peculiar to all madhouses". In June 1816 Thomas Monro, Principal Physician resigned as a result of scandal when he was accused of 'wanting in humanity' towards his patients. Coined in the 18th century, the term 'mad-doctor' referred to a physician treating mentally ill patients of the day. Four generations of one family formed a dynasty of these 'mad-doctors' at Bethlem Royal Hospital between 1728 and 1855. They were the Monros. Bethlem is the world's oldest institution caring for people with mental illness. The Monros' reign of over 120 years, with the 'throne of folly', being passed from father to son, is unparalleled in the history of English hospitals. All four Monros became Fellows of the Royal College of Physicians. Yet the period in which the Monros were physicians at Bethlem, when the treatment of 'the mad' was characterized by bloodletting, vomiting and purging, was plagued with scandals of neglect and brutality. By the early 17th century Bethlem had become the main 'madhouse' in England. But it had also become a byword for madness itself Bedlam. Bedlam has often been portrayed as a place of degradation, with chained and naked patients housed in dark cells and subjected to the gaze of the general public who flocked to Bethlem - one of London's great spectacles. Views of mental illness changed over time. By the

18th century there was a growing concern for patients. They were no longer thought of as objects of curiosity and fun – the view that had prevailed in the preceding century. This led to the scandalous Parliamentary enquiry into 'madhouses' of 1815/16. But by the 18th century Bethlem had also lost its monopoly, and was being challenged by rival institutions. What were the conditions like at other institutions in the 18th and early 19th century? Were the accusations of the Monros brutality and negligence true in comparison to other practitioners of mental health care at the time? Bethlem has been a part of London since 1247, first as a priory for the sisters and brethren of the order of the Star of Bethlehem. Its first site was in Bishopsgate Street (where Liverpool Street station now stands). In 1330 it is mentioned as a hospital, and it admitted the mentally ill from 1377, though by 1403 there were only nine inmates. Early sixteenth century maps show Bedlam, next to Bishopsgate, as a courtyard with a few stone buildings, a church and a garden. Conditions were consistently dreadful, and the care amounted to little more than restraint. There were 31 patients and the noise was "so hideous, so great; that they are more able to drive a man that hath his wits rather out of them'". Violent or dangerous patients were manacled and chained to the floor or wall. Some were allowed to leave, and licensed to beg. It was a Royal hospital, but controlled by the City of London after 1557, but managed by the Governors of Bridewell. Day to day management was in the hands of a Keeper, who received payment for each patient from their parish, livery company, or relatives. In 1598 an inspection showed neglect; the Great Vault (cesspit) badly needed emptying, and the kitchen drains needed replacing. There were 20 patients there, one of whom had been there over 25 years.

Brutal ill-treatment

Bethlem Royal Hospital became famous and infamous for the meted out to the insane. In 1675 Bedlam moved to new buildings in Moorfields, outside the City boundary. In the 18th century people used to go there to see the lunatics. For a penny one could peer into their cells, view the freaks of the "show of

Bethlehem" and laugh at their antics, generally of a sexual nature or violent fights. Entry was free on the first Tuesday of the month. Visitors were permitted to bring long sticks with which to poke and enrage the inmates. In 1814, there were 96,000 such visits. The lunatics were first called "patients" in 1700, and "curable" and "incurable" wards were opened in 1725-34. Eighteenth century Bethlem was most notably portrayed in a scene from William Hogarth's A Rake's Progress (1735), the story of a rich merchant's son whose immoral living causes him to end up in a ward at Bethlem. This reflects the view of the time that madness was a result of moral weakness, leading to 'moral insanity' to be used as a common diagnosis. In 1815, Bedlam was moved to St George's Fields, Lambeth (into buildings - designed by Sydney Smirke - now used to house the Imperial War Museum), where the inmates were finally referred to as "unfortunates." This building had a remarkable library as an annexe which was well frequented. Although the sexes were separated, in the evenings, those capable of appreciating music could dance together in the great ballroom. In the chapel the sexes were separated by a curtain. Finally, in 1930, the hospital was moved to an outer suburb of London, on the site of Monks Orchard House between Eden Park, Beckenham and Shirley. In the early modern period it was widely believed that patients discharged from Bethlem Hospital were licensed to beg. They were known as Abraham-men or "Tom O' Bedlam". They usually wore a tin plate on their arm as a badge and were also known as Bedlamers, Bedlamites, or Bedlam Beggars. In William Shakespeare's King Lear, the Earl of Gloucester's son Edgar takes the role of a Bedlam Beggar in order to remain in England unnoticed after banishment. Whether any were ever licensed is uncertain. There were probably far more who claimed falsely to have been inmates than were ever admitted to the hospital. Richard Dadd, an accomplished artist, who was committed to Bethlem in 1843 after murdering his father following a trip to Egypt, during which he became mentally ill. He was transferred to the newly built asylum at Broadmoor in 1864 and continued to paint, dying there in 1886. The exhibition displays some of his work, which is on loan from The Bethlem Royal Hospital Archives and Museum.

Paupers and Lunatics

By the eighteenth century lunatics were confined in a number of ways: pauper lunatics under the poor law, criminal lunatics under the criminal law, and vagrant lunatics under the vagrancy laws, and others either in profit-making private madhouses or by other private arrangements. The County Asylum Act 1808 authorized justices of the peace to erect and inspect county asylums for pauper lunatics. The crown's jurisdiction over idiots and lunatics in England and Wales with property was exercised by the Lord Chancellor through the masters in Chancery, to whom commissions de lunatico inquirendo were normally addressed, and by Chancery visitors. In 1828 justices of the peace visiting county asylums were obliged by statute to send annual returns of admissions, discharges and deaths to the Home Secretary, who acquired powers to send any visitor he chose to inspect any asylum. The same year powers of inspecting private madhouses and subscription hospitals in the metropolis, except the Bethlem Hospital, were given to 15 Metropolitan Commissioners in Lunacy appointed by the Home Secretary and reporting annually to him. At the same time visiting justices in the provinces were given powers to visit houses there and report to him. In 1832 the statute of 1828 was replaced by another which entrusted the appointment of the Metropolitan Commissioners to the Lord Chancellor, with legally-qualified commissioners being added to the existing medically-qualified ones. In 1842 the Metropolitan Commissioners were charged with carrying out a national tour of inspection of all asylums and producing a report, which they did in 1844. Meanwhile the machinery for dealing with Chancery lunatics was altered by the creation in 1842 of two permanent Commissioners in Lunacy to decide questions of sanity in place of the Chancery Masters. The Lunacy Act 1845 provided for the appointment of 11 new Commissioners in Lunacy, performing the functions of the former Metropolitan Commissioners in Lunacy (whose title however continued in use for some time) but now covering the whole of England and Wales. Their duties involved the inspection and regulation of asylums and the visiting of lunatics, with the exception of Chancery lunatics, who remained under the jurisdiction of the Masters in Lunacy supported by Chancery visitors. The commissioners were responsible for the licensing of asylums in London, but those in the provinces continued to be licensed by JPs. The Lord Chancellor and the Home Secretary shared certain supervisory powers over the commissioners, but the Home Secretary remained completely responsible for criminal and dangerous lunatics. The state institution constructed for them at Broadmoor in 1864 was under his control, with the commissioners reporting to him annually on the condition of the patients.

How did 'bedlam' come to mean uproar and confusion?

It is most probable, and widely believed that over the years there has been a verbal corruption of Bethlehem to Bethlem to Bedlam. The word was used in the 1500s by William Tyndale to mean a madman. In 1247, just outside the walls of the city, the monks from the order of St Mary of Bethlehem founded a priory on land given to them by merchant Simon FitzMary. It is thought that around the early 1300s the poor and the sick were being cared for there – Bethlehem being Hebrew for "house of bread". In1330 it was granted a license to collect alms as a hospital. By 1405 it's known that the 'insane' were being cared for there, through a reference in a Royal Commission. From this caring start it would appear that from around the middle of the 16th century, when it was handed over to the City of London, some of the Keepers (managers) were somewhat patient-care less orientated. There are descriptions of the inmates making a terrible noise described by one source as "so hideous, so great that they are more able to drive a man that hath his wits rather out of them" and of patients being restrained with chains. Even when it moved to new premise in1675 things didn't get any better, and by the 18th century it became a form of entertainment to visit the asylum to view the poor wretches. One source quotes that the visitors could "laugh at their antics, generally of a sexual nature or violent fights" or even poke them with sticks. The financial accounts show that in one year alone visitors generated £400 at one penny per visitor. Pre decimalization, there were 240 pennies in the British pound, so that equates to 96,000 visitors in one year!

Bedlam - The Hospital of St Mary of Bethlehem

Bethlem Royal Hospital is a psychiatric institution in Greater London. Despite its Kent postal address, it is in the London Borough of Bromley. It currently treats people with mental health and substance misuse problems. Services also include specialist units available for people across the UK and local psychiatric services for people from the Borough of Croydon. There are more than 1700 patients admitted each year, and over 50% of the patients are cured. It has over 800 members of staff. This, however, is not the story of the current hospital; rather the story of Bedlam, where the Bethlem Royal Hospital began over 700 years ago. Bethlem and Bedlam are two of the many medieval variants of Bethlehem, and are used throughout this entry. From as early as 1329 it was being used as a hospice, and received a licence to collect alms for the poor and needy of the local area and the Mother Church in Palestine. In 1375 Bedlam became a royal hospital. It had been taken by the Crown on the pretext that it was an alien priory[2]. It began the care of mentally ill patients two years later (the word 'care' here means chained to the wall and, when violent, whipped or ducked in water) and specialised in the care of mental illness from 1403. Records show that at this time it had one master, a porter and his wife and nine inmates. There were a number of servants. Patient numbers steadily increased, and records show that it was being called 'Bedlam' by 1450. Early 16th century maps show 'Bedlam Gate' next to Bishopsgate. It led into a courtyard where a few small stone buildings were, a church and a garden could also be found here. There were 31 patients housed in a space for 24, and the noise was 'so hideous, so great; that they are more able to drive a man that hath his wits rather out of them'. The whip and chains were the usual 'treatment' for patients. Some patients were allowed to leave the 'madman's pound'. They were 'tagged' with a tin badge on their left arm to signify their status. They were called 'antics' or 'God's minstrels' because of their habit of roaming aimlessly. Violent or dangerous patients remained in Bedlam, manacled and chained to the floor or wall. The main ward was called Abraham - an Abraham man was a beggar who pretended to be mad.

Bishopsgate Henry VIII

Henry VIII granted Bethlem a charter as a hospital for the insane in 1547 when the priory was dissolved. The government of the hospital was granted to the City of London (with all its revenue), on condition that the City spend a certain amount on new buildings. It became a City institution as well as a Royal hospital. The management was granted to the City in 1557, who transferred it to Governors of Bridewell. Bridewell had originally been a palace, but its position next to the stinking Fleet River meant that royalty soon deserted it. It became a hospital, but was mainly used as a prison. The Governors left the management of Bedlam to keepers (the replacement for the old 'masters'), who made what they could out of it in the same way that the keepers who ran prisons did. Each patient was paid for by his or her parish, livery company, or relatives. The charges largely depended on what the keeper thought the market could bear and what he thought he could get away with. At this time it was the only fee-paying, specialised hospital in London. The Governors gave most of their attention to Bridewell, and the patients of Bedlam suffered terrible neglect. In 1598, Bridewell Governors inspected Bedlam - nothing had been spent on upkeep of the buildings, the 'Great Vault' (cesspit) badly needed emptying, and kitchen drains and sinks needed replacing. At this time there were 20 patients there - one who had been sent from Bridewell over 25 years previously. Five others had joined him from Bridewell over the last decade. Records do not show if any improvements were made at the time of this inspection. We can only hope that there were. By early 17th century Bedlam was the only hospital for the insane in the country. Most of the patients were vagrants, apprentices and servants. There were a few scholars and gentlemen. Half a century later Bedlam had become badly over-crowded, noisy and polluted.

Tales of Bedlam

Bedlam, Bethlem Royal Hospital, must surely be one of the most famous hospitals in the world. It's been around, one way or another, since 1247 and is infamous as a lunatic asylum. The history of the hospital itself against the history of the ways in which the mad were treated, and against what she sees as a rising tide of madness within society. The lunatics were first called "patients" in 1700, and "curable" and "incurable" wards were opened in 1725-34. In the 18th century people used to go to Bedlam to stare at the lunatics. For a penny one could peer into their cells, view the freaks of the "show of Bethlehem" and laugh at their antics, generally of a sexual nature or violent fights. Entry was free on the first Tuesday of the month. Visitors were permitted to bring long sticks with which to poke and enrage the inmates. In 1814 alone, there were 96,000 such visits. 'It was so loathsomely and filthily kept that it was not fit for any man or woman to come into. Situated variously in Bishopsgate, Moorfields and Lambeth, one of the main attractions over the centuries for the London mob was the Bethlehem Royal Hospital or Bedlam'. So famous has the hospital become that the word has been accepted into the English language signifying 'a scene of wild uproar'. The lunatic asylum made a lot of money from the public up to the year 1770, as visitors were admitted to see the lunatics as we might visit the zoo today, the entrance fee being 1d. In a report to the House of Commons in 1815. Dr. Connoly reported that he found in one of the side-rooms; "about ten patients each chained by one arm or leg to the wall, each wearing a sort of dressing gown with nothing to fasten it. Some sensible and accomplished, some imbeciles. Many women were locked up naked with only one blanket." When finally the date of her release arrived she became tranquil, nursing two dolls which she imagined were her children. Another patient, well-known to the many visitors, wore a straw cap and promised to declare war on the stars if rewarded with a bottle of wine. One of the most famous patients, often visited by members of Parliament was a certain William Morris. For twelve years he was chained with a strong iron ring round his neck His arms were pinioned by an iron bar and he could only move twelve feet away from the wall. In this position he lived as normal a life as possible before dying shortly after his release. Two more patients spent a total of over eighty years between them in Bedlam for trying to kill the same man. James Hadfield was confined for 39 years for attempting to shoot George III. He spent his time writing verses on the deaths of his cats and birds, his only companions in the hospital. Margaret Nicholson spent 42 years in solitary confinement for attempting to stab the same King. William Cooper described his thoughts on visiting the asylum as a youngster. "The madness of some of them has such a humorous air, and displayed itself in so many whimsical freaks, that it was impossible not to be entertained at the same time that I was angry with myself for being so." The life stories of some of the patients who finished their days in Bedlam make fascinating reading. Hannah Hyson died within days of being rescued by her father from Bethlem, her body covered in scabs and her knuckles red raw where she had crawled about her cell on her hands and knees. Ann Morley, a former patient at Bethlem, was admitted to Northampton Asylum in a skeletally weak condition, incontinent, prolapsed and close to death. Upon recovery, she testified to being punched in the face by a bad-tempered nurse called Black Sall (the name referred to Sall's moods), hosed down with freezing water and being made to sleep naked on straw in a cellar. It was only with the arrival of William Charles Hood, in 1853, that Bethlem began its long process of reform, and even after this date episodes of cruelty and neglect surfaced, with a high suicide rate attracting press coverage in the 1880s. By the turn of the century, Bethlem had undergone a transformation: pauper lunatics had been banished to the great asylums on the fringes of London; the worried well and the shabby genteel, driven to madness by the pressures of middle-class life, inhabited a comfortable asylum that appeared, at first glance, more like a Pall Mall club than a psychiatric institution. In 1930, the hospital was relocated to Kent, while the imposing Victorian building in Southwark, with its distinctive pumpkin-shaped dome, took on a new role as the Imperial War Museum. Today the hospital, where comedian Jo Brand was a psychiatric nurse, looks more like a series of villas. There, ancient books, the size of broadsheet newspapers, list the case histories of former patients.

Richard Dadd

The renounced Artist

For many years, Dadd has been perhaps the most celebrated Victorian resident of Broadmoor. An artist of some repute, the quality of his fairy paintings was acknowledged during his lifetime, and he continued to paint remarkable works during his time in asylums. Many of these works survive, and quite apart from any sensational interest in Dadd's circumstances, it is acknowledged that Dadd possessed a rare talent.

His artistic endeavors had benefited from conducive surroundings. Dadd's father, Robert, was an intellectual man, a chemist and the first curator of the Chatham and Rochester Literary and Philosophical Institution's museum, and Dadd himself attended The King's School at Rochester. When he was seventeen, the family moved to London, and at nineteen he was admitted to the Royal Academy Schools where he completed his training as an artist.

Dadd had been born on 1st August 1817 in Chatham. He was the fourth of seven children borne by his mother Mary, a total of four of whom would eventually die insane. The young Dadd was influenced by both literary and classical themes, and by the early 1840s had begun to create the fairy paintings for which he would become best known. In due course, his work attracted the patronage of Sir Thomas Phillips, a solicitor from South Wales who had been knighted for his part in ending a Chartist riot, and who had money to burn. Phillips decided that he wished to undertake the Grand Tour of classical sites across Europe, and he recruited Dadd to accompany him as his personal artist, and draw what they saw.

They began their journey in July 1842, traveling first through Belgium, Germany and Switzerland before reaching Italy, then moving on to Greece, Turkey and Palestine. Dadd seemed to enjoy the tour, and wrote various letters home detailing his wonderful experiences. He was fascinated both by the scenery he encountered and the people he met, and an internal record of these compositions appears to have remained locked within him during his years of treatment. Decades later he would bring it out to influence the works he completed in Bethlem and Broadmoor.

Although the tour itself was meeting expectations, by the time they reached Egypt Dadd had begun to exhibit signs of mental illness. His mind had been unfettered and was running free across a new spiritual landscape. He was soaking up the culture of ancient Egypt, and this was to influence his belief structure for the rest of his life. Dadd appears to have been aware of his increasingly weak grip on reality, and some of his letters hint at an effort to try and rationalize the source of his feelings. His health seems to have deteriorated very quickly from this point. He and Phillips crossed to Malta and then to Italy again, and by now he was reporting regular delusions. He would later describe his first irrational impulse to the staff at Bethlem and Broadmoor - his desire to kill the Pope at a public appearance in Rome, an impulse he resisted as he felt the Pope was too well protected.

When they reached Paris again in May 1843, Sir Thomas sent for a doctor to examine his traveling companion, and Dadd was duly sent home. Phillips wrote to the family that Dadd's character had completely changed to becoming that of a suspicious and withdrawn man. Over the summer, friends and family became increasingly worried about and wary of him. A doctor consulted by the family recommended that Dadd was committed to a private asylum and put under immediate restraint. His father was reluctant to agree. This was to be the last act before Dadd took matters into his own hands. On 28th August 1843, Dadd asked his father to accompany him to an inn at Cobham, near Gravesend in Kent. After enjoying a meal together, they walked to nearby Cobham Park, where Dadd attacked and killed his father, first trying to cut his throat with a razor, and finally stabbing him with a knife.

Dadd was aware that he had done something wrong, even if he was not exactly sure who or what it was that he had killed. He fled to France. He later stated that he was on his way to kill the Emperor of Austria, but whatever the truth in that, only two days after killing his father he attacked a complete stranger who was his traveling companion, while riding in a carriage through a French forest. He was arrested by the French authorities and identified himself as a wanted man for the Cobham killing.

Initially, Dadd was sent to a succession of French asylums, having been certified insane, before he was extradited to England in July 1844. He never stood trial for the murder of his father, and was found insane when he came to plead. He was duly given the HMP order - to be detained at Her Majesty's Pleasure - and sent to the criminal lunatic ward at Bethlem on 22nd August 1844. Like for Oxford, records of Dadd's time at Bethlem are available at the Bethlem Royal Hospital archives.

It was only after he had been received into custody that some explanation came as to Dadd's motive for his acts. When he was arrested in France, the police had found on him a list of 'people who must die', with his father's name at the top. A search of his lodgings in England had uncovered various portraits of his friends, all with a bloody slash from the artist added across their throats. Notes from his stay at Clermont Asylum in France indicated that Dadd believed that his father was the devil, and that the son had been commanded by the Egyptian god Osiris to kill both Robert Dadd and other people. This was a delusion that Dadd maintained once he was in Bethlem. He remained convinced that he was on a mission to battle the devil, who could take many forms, including that of Dadd senior, and that the artist formerly known as Richard Dadd was in fact descended from Osiris.

Almost immediately that he was confined to Bethlem, Dadd began to paint again, something that; happily, he would continue to do for the next forty years. He appears to have been very insular during his time at Bethlem, and did not associate much with other patients. However, he formed a close bond with the man who was both his and Edward Oxford's doctor, Charles Hood, the superintendent of Bethlem, and also with steward George Haydon, the same steward that Oxford had apparently communicated with years after his own discharge. Hood was a reformer who sought to create Bethlem as a refuge for its patients, in the modern Victorian fashion. Haydon was a writer and artist as well as being a steward of lunatic asylums. It seems reasonable to assume that Dadd found some encouragement from them both.

As in Oxford's case, the Broadmoor case notes repeat some observations on the patient made at Bethlem in 1854: 'For some years after his admission he was considered a violent and dangerous patient for he would jump up and strike a violent blow without any aggravation and then beg pardon for the deed.

This arose from some vague idea that filled his mind, and still does so to a certain extent, that certain spirits have the power of possessing a mans body and compelling him to adopt a particular course whether he will or no.' He was reported to binge eat until he vomited, and otherwise behave eccentrically, believing that he was possessed of special powers.

Also like Oxford, Dadd was amongst the tranche of Bethlem patients who were transferred to Broadmoor when the latter opened. Dadd made the great trek to the Berkshire countryside on 23rd July 1864, a few days short of his 47th birthday. At the prescribed initial examination of a new patient, his tongue was recorded as being 'broad and flabby'. He was also still convinced of his delusions, believing himself to be a marked man 'under the influence of an evil spirit': 'Makes laboured attempts at justification of the two criminal assaults saying it was in "justification of the Deity".'

He settled in to his new accommodation quickly, and soon began painting again. By November 1864 his case notes record that he was engaged in a detailed fairy painting. He received money from his family regularly, and in the patients' account books kept by the Hospital his careful signature records his receipt of brushes and board that he purchased for his work. These accounts also record many purchases of foods with strong flavors, such as herrings, gingerbread and peppermints. Patients were allowed to maintain funds for their own use, and trusted patients such as Dadd made good use of this concession.

For Dadd was a tranquil patient, whose madness only became apparent during conversation. His notes regularly state his seeming contentment, as well as the continuation of his delusions. One conversation with Dadd written up by William Orange was on the subject of chess, and how some people possessed a spirit that allowed them to play chess 'without the board'. Dadd further mused that chess pieces could be unfriendly towards some players due to the 'antiquity of the game'. Evidently nothing could escape the ancient pull of Osiris.

Dadd suffered from gout from time to time, though was also able to keep up an intake of wine and spirits, and suffered a prolonged bout of illness during 1868-1870. By 1870, he was recorded as having lost three stone over the past two years. However, he had recovered sufficiently by 1872 to begin to paint decorations around the stage in Broadmoor's Central Hall, which he continued for several years. He also completed his well-known 'Portrait of a Broadmoor Officer' in 1875; long supposed to be a likeness of Dr Orange, it is more likely to be of Dadd's twenty-nine year-old doctor and Orange's deputy at the time, David McKay Cassidy. Nor were these Dadd's only contribution to the furnishings of the estate: Orange's son also remembered the artist painting a mural along one wall in the Medical Superintendent's house, work which, like most of the Hall decorations, is now lost.

In 1877, there is the only note made at Broadmoor relating to Dadd's reason for admission. David Nicolson recorded a detailed conversation that he had had with Dadd about the murder of Dadd's father. Dadd stated that he was not convinced that the man he killed was his father, presumably clinging to the belief that he had instead attacked the Devil. Rather, Dadd had been convinced at the time of the killing that the 'gods and spirits above' required him to make a sacrifice. Dadd was able to describe the murder scene and his reaction when his father fell. Nicolson wrote: 'Dadd (posing himself with upstretched arm), thus apostrophized the starry bodies "Go," said he "and tell the great god Osiris that I have done the deed which is to set him free."' Dadd also stated that the attack in France had been brought about by his observation that two stars in the constellation of Ursa Major were moving closer together, thus convincing him that a further sacrifice was demanded by the ancient gods.

Despite his continuing delusions, Dadd was evidently no bother to the medical authorities. He remained insane, but in other respects simply became another old man, occasionally wandering about the grounds to watch the other patients playing cricket. His disappearance underneath the Asylum radar is evidenced by the fact that no entries were made on his case for a whole seven years, from 1878 until 1885, at which point he was removed to the infirmary in Broadmoor's Block 3 with what proved to be his final illness. It was back to where Dadd had spent his first years in Broadmoor. There is evidence to suggest that he was later moved to Block 2, the 'privilege' block, where the better behaved patients were allowed more freedom, as he appears to have been observed in a room there by a journalist touring the Asylum in the early 1880s.

Wherever he spent most of his days, he stayed in the infirmary from June 1885 until his death on the evening of 8th January 1886, aged 68, from tuberculosis. The end had been quite quick, with Dadd still getting up and about until a week before he died. He was buried at Broadmoor. In common with a significant proportion of Asylum patients, he had outlived most of his immediate family, and there were no immediate relatives left to mourn his passing. There are a few papers in his Broadmoor file which relate to the dispersal of his estate, though the file adds that various letters from solicitors had been taken from it by the Broadmoor steward, sadly never to be returned.

Dadd's reputation was recognised during his lifetime, though due to his situation he was not particularly celebrated and only rarely exhibited. His passing was not noted at the time of his death, and it was only in 1974 that the first major exhibition of his work was curated, at The Tate. A substantial collection of his work is held at the Bethlem Royal Hospital Museum in south east London, which includes a number of paintings that remained at Broadmoor after Dadd's death. One of his most significant works from Bethlem, The Fairy Feller's Master-Stroke, is now on permanent display at Tate Britain in London. A lost work, The Artists' Halt in the Desert, was discovered in 1987 during filming for the BBC's Antiques Roadshow and is now in The British Museum. Interest in Dadd's work only appears to deepen with time, and there seems little chance that this particular Victorian artist will ever be forgotten.

William Chester Minor

Probably Dadd's rival for the crown of best-known Victorian Broadmoorite is Dr Minor, American medic, murderer and contributor to the first Oxford English Dictionary. Minor was the subject of Simon Winchester's best-selling book The Surgeon of Crowthorne, which is an entertaining and thorough account of his life, and should be easily found by anyone wishing to explore Minor's story in greater detail.

Winchester records Minor's birth as having been in June 1834 in Ceylon, now Sri Lanka. He was the son of missionaries, and one of two children. His mother died of tuberculosis when he was three, and his father subsequently remarried and had a second family. Minor remained in the east with this extended brood until his father sent him to live with his uncle in New Haven, Connecticut, when Minor reached the age of fourteen.

Once in the US, he attended Yale University, where he studied medicine. He graduated in 1863, and joined the Union Army as a surgeon in the middle of the American Civil War. Winchester says that Minor was sent into action at the awful and bloody Battle of the Wilderness in May 1864, and that this experience haunted him. At Minor's trial, years later, his defense suggested that the horrors of war had caused his mental illness. Particularly, he had witnessed an execution, and had been required to brand an Irish deserter from the Union cause with a letter 'D'. Whilst this theory will have to remain conjecture, it presents a powerful picture of a traumatized individual, which Minor certainly was.

After the end of the civil war, Minor remained in the American army and indeed rose through the ranks. The pressures of his work continued, though without him showing any immediate signs of insanity. The only catalyst presented for the change in his behavior is hearsay: that he had become engaged, but that the relationship ended. It is the earliest point in Minor's story that sex enters the narrative, though it seems unlikely that Minor had not already been consumed by sexual thoughts before this point. What is known is that he was discovered frequenting brothels in New York, where he was stationed at the time. Such behavior might be considered normal for a soldier, even tacitly encouraged, but instead there must have been something about Minor's behavior that was not normal. Bearing in mind his subsequent history, the possibility that Minor was engaging in either homosexual or bisexual acts might be one possible conclusion. A deliberate move was organized for Minor to Florida to remove him from a scene of temptation, but this failed when he began to exhibit delusions of persecution by his fellow officers. In 1868, the army diagnosed him as suffering from the mental illness of monomania, or an obsession with one subject, which gave rise to delusions. He was sent to the Government Hospital for the Insane in Washington DC (now St Elizabeth's Hospital).

Despite his obviously continuing illness, Minor was released from St Elizabeth's in 1871, though now a man in enforced retirement from the army and also in receipt of his pension, which he could add to an income from his well-to-do family. He traveled to London at the end of the year, ostensibly to spend time touring Europe. He did not make it any further. It appears that he first took up residence at Radley's Hotel, in the West End, before moving to Lambeth after Christmas, where it seems likely he felt he would have easier access to the sex trade. It was in Lambeth that he shot and killed a stranger called George Merritt or Merrett on 17th February 1872. Minor had already approached Scotland Yard, reporting that he was being followed and otherwise persecuted by various nameless men. The warning was ignored. One night Minor woke, and saw a figure at the end of the bed which he reckoned to be one of his abusers. He pursued the phantom spirit into the street, where Minor chanced upon Merritt walking to work at a brewery near to Waterloo. Merritt was married and had six children, with another on the way, and that night he was simply in the wrong place at the wrong time. Minor chased him, pursued him as he ran, and then caught and shot at him several times before fatally wounding him in the neck.

The scene of crime was very central, between Waterloo and Hungerford bridges, and Minor was apprehended on the spot. Minor said it was a case of mistaken identity that he had thought Merritt was a person who had been breaking into his room. While the mistake was fleeting, the intention was permanent, and the delusion about needing to fight forced entry to both his room and his person would remain with Minor for the rest of his life.

Minor was committed for trial, and this was held at the Surrey Assizes in Kingston upon Thames in April 1872. The nature of Minor's enduring delusion was laid bare at the trial. A warder at the jail where Minor was on remand was also an employee at Bethlem, and he testified that every morning Minor would wake up and level the accusation that his guards had allowed him to be sexually abused during the night. His abusers hid in the voids of the room – under the bed, or in the walls or rafters. The abusers were always male, but both men and women (and boys and girls) feature in Minor's later descriptions of the sexual terrors that his abusers forced upon him. Minor's step-brother attended the trial to confirm that this delusion could be dated back to at least his release from the Washington asylum. Minor would frequently report that people had been in his room at night. This was the subject of his monomania.

His step-brother stated that apparently it was all punishment for an unspecified act that Minor had been forced to commit while in the Union Army.

Whatever Minor's confused reasoning for his actions, the jury was quite clear that he was not guilty by reason of insanity. He duly received the sentence of detention at Her Majesty's Pleasure, and was sent on to Broadmoor.

Minor arrived from the Surrey County Gaol on 17th April 1872. Unusually for a Broadmoor patient, he traveled with another patient being transferred from the same prison, a gentleman called Edmund Dainty, who had killed a fellow patient in the Surrey Asylum. Described on admission as 'A thin, pale and sharp-featured man with light coloured sandy hair; deep-set eyes and prominent cheek bones', Minor dutifully recounted his persistent nocturnal experiences, as well as giving an account of his current bodily health (gonorrhea and possibly signs of tuberculosis, though none were found). Like Dadd, his delusions appeared to be self-contained and manageable, and he was obviously thought to be a low risk and was placed in Block 2, where privileges were greatest.

Minor was one of a small band of foreign nationals in Broadmoor, though most of these had become naturalized even if they were not citizens, and they did not quite have the character of a tourist that Minor's case suggested. As a result, almost as soon as he arrived in the Asylum, the American Consulate in London wrote to Dr Orange for permission to send various things to Minor – both his own possessions and 'some comforts, such as Dunn's Coffee, French Plums etc'. The Consulate sent on Minor's retrieved possessions shortly after, including clothes, drawing equipment, his tobacco and his diary. They kept hold of his surgical instruments, which had also been found in his rooms.

As a patient in Block 2, Minor enjoyed a reasonable degree of freedom within the Hospital routine. He had his own clothes, his art materials, and a regular income from his family which allowed him, like Dadd, to ask the Hospital to purchase things for him. Examples of things Minor bought include: beef, haddock, poultry, game, steak, bacon, salmon, as well as biscuits, coffee and lots of eggs. Once he bought himself a macaroni cheese. He also regularly purchased newspapers and a number of engineering journals (quite possibly for advice about solid building construction, which might prevent his nightly suffering).

He experienced as comfortable an existence as would be possible for any Broadmoor patient. At some point, he was allowed a separate day room as well as his bedroom, where he presumably kept his books, and by 1901 if not before he employed another patient as his servant (occasionally having to change his domestic staff if they were discharged). Winchester suggests that Minor's two rooms were interconnecting, but this is unlikely as he was a tenant rather than a freeholder, and more probably they were either next door to each other or close together in Block 2. The exact date from when the extra room was granted is not clear, though it is likely that it postdates 1876, when Orange succeeded in having most of the convict patients transferred to Woking Prison. Certainly, Minor must have enjoyed the privilege for most of his stay, as a note in his file from 1887 suggests that Minor could not get into his day room one morning as the lock was faulty (which no doubt provided him with further evidence of the conspiracy against him), until the attendants had removed an obstruction from it.

Much of the anecdotal evidence for Minor's comfort comes from a 1958 letter written by Dr Patrick McGrath, then the Superintendent, in response to an academic enquiry. He reported on a conversation with the daughter of David Nicolson, Superintendent 1886-1895, who confirmed most of the details above. Miss Nicolson also reported that as well as his own library, music and paints, Minor had a private stock of wines and spirits, played the flute, and would from time to time dine with the Superintendent's family in the latter's home.

Minor was obviously cared about by his family and friends, and received regular visits as well as money and luxuries. With cash to spend and time to kill, he began to amass books and read voraciously. After Sir James Murray published his 'appeal to English speakers and the English reading public' in 1879 for help with what became the Oxford English Dictionary, Minor must have come across it in his newspapers and felt a call. He began immediately to send in to the dictionary staff what became thousands of examples of word use from his book collection to assist them with their Herculean labours.

Books would come to play a part in the refinement of his delusion. In his early years in Broadmoor, he was convinced that poison was administered to him at night. Usually chloroform was used to render him helpless to abuse and humiliation. By 1877, this had changed to his being subject to torture by electricity, and by 1878 he was being secretly removed from the asylum at night and abused. All these actions were evidently attacks upon his free will. Once they had his body, the next sacred thing in line were his books, and the first evidence that the criminal agents had moved on to these dates from 1884, when he wrote to the Superintendent alleging that items in his library were defaced at night.

Minor must have found the approach of night a very frightening thing, as it brought with it the certainty of pain and degradation. Immediately that he arrived in Broadmoor, he would barricade his room every night by placing furniture across the door of it. Only very occasionally would the attendants report that his nights had not been restless; usually, the morning brought fresh reports of his sordid trials. He expended much effort on trying to remedy the situation through practical means such as the barricade, asking the Superintendent to keep a close watch on the attendants and so on. He was also always open to offering other solutions. The letter below was sent to Orange on 6th October 1884:

Dear Sir

Let me mention one fact that falls in with my hypothesis. So many fires have occurred in the US originating quite inexplicably in the interspace of ceiling and floor; that I learn now Insurance Companies refuse to insure large buildings – mills, factories etc – which have the usual hollow spacing under the floor. They insist upon solid floors. All this has come to notice within ten years; but no one suggests any explanation.

Very sincerely yours

WC Minor

Amongst the more interesting discoveries in Winchester's book is the suggestion that Minor also met regularly with Eliza Merritt, the widow of the man he shot. Unfortunately nothing has yet surfaced in the Broadmoor archives to verify this. However, we do know that through Minor's work on the Dictionary, he met with Sir James Murray. Indeed an apocryphal account of the meeting has been around for some time, the story being that Murray was received into Dr Nicolson's office, then the Medical Superintendent, whereupon Murray thanked Nicolson for his contribution to the dictionary. Nicolson corrected Murray and assured him that it was not he that should be thanked, and then walked him to Block 2, through the corridors of howling lunatics (or at least, painting and reading lunatics) and introduced him to Minor. Murray's reaction was to gasp through his generous beard in amazement.

In reality, Murray knew who and what he was visiting before he made the journey down from Oxford. Beyond that, the extent of the relationship between the two men is open to conjecture. Evidence from Minor's file suggests that they met sporadically. The first letter from Murray in Minor's file is dated 3rd January 1891. It refers to Thomas Brushfield, a former Superintendent of Brookwood Asylum in Woking and probably a contemporary of Dr Nicolson. Murray wrote that he was currently working on 'do' for the Dictionary, and wished to make arrangements to visit Minor for the first time. Whether or not he became a regular visitor is not evidenced in Minor's file, though the next letter in the file from Murray, which is dated 21st August 1901, says that Murray had not seen Minor since just before Dr Nicolson left as Super. That places their last previous meeting as towards the end of 1895, and implies that at the time of writing, Murray had not visited Minor for six years.

Despite the therapeutic effects of his work on the dictionary, Minor's condition deteriorated over the years. The delusions and frustrations never left him. Reading his notes gives a sense that sometimes he probably internalised them, and that when it all got too much he would suddenly explode, making an accusatory outburst to the attendants or to the Super. Eventually he took matters into his own hands, and on the morning of 3rd December 1902 he tied a tourniquet around the base of his penis and sliced off the offending organ. He was 68 years old, and had never been able to come to terms with his own sexual urges. Asked why he had done it, he replied: 'In the interests of morality'. He testified that for a long time previously he had been taken out of the asylum at night and forced to fornicate with between fifty and one hundred women 'from Reading to Land's End.' He spent time in the infirmary but was discharged after four months back to Block 2. Sadly of course, his retaliatory act did not defeat his delusions, which remained as before. In his last letter to Dr Brayn, shortly before his discharge, Minor was complaining still of 'these nightly sensual uses of my body that I experience and struggle against.'

As indicated earlier, the nature of these 'sensual uses' may provide some help in understanding Minor and his mental illness. Winchester's book suggests various hypotheses about Minor's own sexual motivations, from dusky eastern maidens with pert breasts to disease and prostitution in New York's metropolis, and to guilt about his feelings for Eliza Merritt. However, Minor's early delusions at Broadmoor all seem to relate to his body being used by men, and it is only in the later years that women play the more significant part. To the modern reader, Minor may be repressing homosexual or even pedophiliac tendencies as much as heterosexual ones. Plenty of things may have happened to Minor before he came to the attention of the authorities, though we may never know exactly what Minor's own sexual experiences were, and how his obsessions led him to such a dramatic conclusion. What is beyond doubt is that Minor was able to concoct outrageous tales of depravity, experienced with a multitude of other bodies, of both sexes and all ages, and that his mutilation of his own body was a direct result of his discomfort with that fact.

Still riddled with fear and hampered by his lifelong burdens, Minor was also becoming a very old and frail man, which brought on additional problems. In December 1907, he neglected to check the water temperature and severely scalded himself when bathing in his room. In 1908, he suffered from a serious bout of flu. The facts of his advancing years and ill health was not lost on his family and friends, who remained in constant touch with the hospital. The first formal petition for Minor's release was delivered to the Home Office in 1899, who rejected it quickly. But by 1903, Dr Brayn was suggesting to Minor's step-brother that a proposal to remove Minor to America might be received favorably, providing suitable care could be found for him.

It took seven years before matters reached a resolution. In 1909 and 1910, Dr Brayn felt compelled from time to time to remove Minor to the infirmary, not thinking it safe to leave Minor alone in his room day after day, as he was no longer capable of looking after himself. Laid up, and deprived of his books and his art materials, Minor was increasingly miserable, as well as increasingly harmless. Finally, in April 1910 a conditional discharge was granted for his release. Both Sir James and Lady Murray visited him one last time before he was escorted to the Tilbury Docks on 15th April (via Bracknell, Waterloo and St Pancras), where he was put on board a steamer and handed over to the care of his step-brother for the journey back across the Atlantic.

After thirty-eight years in Broadmoor, Minor arrived back in America to return to the Government Hospital for the Insane in Washington. There he swapped one similar regime for another: a private room, certain privileges, and nightly torments. Though the Broadmoor authorities had thought he was nearing the end of his life, he did in fact keep going for a further decade, reading, writing, and making the occasional outburst. He remained in Washington until November 1919, when he was compassionately released to be nearer his family, at the Retreat for the Elderly Insane in Hartford, Connecticut. He died there on 26th March 1920.

Inevitably for Minor there has to be a postscript, because unlike Dadd, whose work was acknowledged during his life, Minor's place in history has only really been secured after his death. Hayden Church, an American journalist and author of the imagined Minor/Murray first meeting, published one romantic piece about Minor in 1915, and another in 1944. He intended to write a book about Minor – there is a relevant letter in Minor's file stating this intention – but eventually did not. Little more happened until the 1980s, when the Oxford University Press was becoming aware of its own history and Minor's place in it, and a more scholarly article about the American in Crowthorne was published. Then came Simon Winchester, a full biography and worldwide recognition. Once it happened, it seemed like it was an obvious conclusion: Minor's story is ultimately one of triumph in adversity, and that always makes for a good read. Revisiting Minor's life for this short piece has made me realise how much might still be written about him, for while aspects of the man might have become obscured by the clouds of myth, there is a man to be discovered, all the same.

Where did the patients come from?

For most of its history, Bethlem's patients have come from the ranks of the very poor, and with a few notable exceptions, little is known of most of them beyond their names and places of origin, recorded in admission registers which date back to 1683. Even the more detailed casebooks of the nineteenth century tell us relatively little of these people outside their brief passage through the hospital, though a remarkable set of photographs taken in the 1850s adds an extra and moving dimension to the record. A change in social status took place in the mid-nineteenth century. It was decided to exclude pauper patients, who could now be cared for in the new county asylums, and give preference to the poor of the middle classes. In 1882 the first handful of paying patients were admitted, and inevitably the numbers crept up (though the 'free list' was never wholly abandoned). By the time the new hospital opened in 1930, the prospectus was referring to "accommodation for 141 ladies and 109 gentlemen, each of whom must be of suitable educational status. Under the NHS, in conjunction with the Maudsley, Bethlem became part of a system based largely on specialist units and treatments, and clinical considerations took over from social and financial circumstances as the main criteria for admission. The new Trust has now taken responsibility for mental healthcare in the Croydon area, introducing for the first time (for Bethlem though not for the Maudsley) a service for the local community. As the 1598 list already shows, patients were admitted from all over the country. This reflects the fact that for several centuries, Bethlem was the only public institution for the mentally disordered. (The main local alternatives which were to grow up later were the private madhouses, which flourished from the eighteenth century on, and the county asylums of the nineteenth century.) Contrary to popular belief, mental disorder has been regarded as potentially curable throughout the hospital's history. Whether popular 'cures' that have been in vogue throughout history have actually contributed to the patients' recovery is another question. However, a proportion - possibly just over 30 per cent in the eighteenth century - did recover, though some were readmitted at a later date. In the absence of effective treatments, most of which have been developed in the present century, physical restraint featured prominently in Bethlem's regime, but this was never the whole story. Both medical and psychological methods of treatment have long been in use.

Medicines have probably always been administered, and in 1700 a fund was set up to provide 'Phisick' for discharged patients to prevent their relapse. However, under the Monro family's long reign as physicians, medication made little progress. Their practice, handed down from father to son, can be summarized as 'purges, vomits, and bloodletting' - standard treatment in the early eighteenth century, but distinctly outmoded by the mid-nineteenth. Cold and warm bathing, introduced in the 1680s, seems more likely to have benefited the patients, particularly 'In the Heat of the Weather... to cool and wash them'. Occupation, and distraction from false, deluded or melancholy thoughts, was also considered important. By 1765, for example, it was one of the matron's duties to make sure that the women patients who were 'low spirited or inclinable to be mopish' should get up and not be allowed to slink back to bed, and to employ those who were capable at needlework. In the 19th century occupation and entertainments multiplied. Dances were held in Bethlem as early as the 1840s, outings and excursions took place, and many other activities were introduced. Huge advances in scientific knowledge, particularly in the understanding of brain function and the development of effective drugs for many conditions, together with a wide range of techniques based on psychotherapy, have revolutionized treatments in the second half of the twentieth century. The Bethlem Royal Hospital continues to play an important role in today's modern NHS. It provides specialist services, including those dealing with prenatal psychiatry, challenging behavior, addictions, psychosis, child and adolescent psychiatry and forensic psychiatry. Here many of the advances in neuro-imaging, genetics, biochemistry and neurophysiology are put into practice, targeting major mental health problems - problems that continue to be a leading cause of illness, distress, disability and mortality. Policy changes in the NHS and Social Services call for the care of the mentally ill to take place in the community; nevertheless, the need for in-patient care continues. There are arguments against the further closure of hospital beds, especially within inner-city areas, and opposite arguments from those who believe the community care programme has not developed enough.

Shell shock sufferers in Bedlam

Foreigners have always known the English were mad. They put it down to the dreary climate, the rigours of nonconformity and the tedium of the English Sunday. Long before that, the gravedigger in Shakespeare's Hamlet spoke of sending the Prince of Denmark to England, where his madness would not be noticed. England was enough to drive people mad, another asserted: 'There you shall see many discontents, common grievances, complaints, poverty, barbarism, beggary, idleness, epicurism - cities decayed, base and poor towns - the people squalid, ugly, uncivil.' Londoners flocked to Bedlam to laugh at the antics of the inmates: a visit to the madhouse was a good day out, ranking with a public execution and featuring in all the popular tourist guides. The freak show at Bedlam was a mirror of the city's disordered psyche. London itself was mad, with law and order on a knife edge, and gambling fever, prostitution and alcoholism rife. It was inevitable that in The Rake's Progress, Hogarth's Tom Rakewell should end up in Bedlam, driven to madness by debauchery and surrounded by a cast of grotesques. Madness was a reflection of the state of the nation. Where better to recruit the nation's politicians, the satirist Jonathan Swift suggested, than Bedlam, since they could not be any more insane than the ones in power. Political cartoonists Gillray and Rowlandson suggested that contemporary politicians belonged in Bedlam and depicted Charles James Fox raving in a straitjacket the standard restraint for the poor inmates of Bedlam, but for the poor inmates of Bedlam, but also for George III, the victim of a rough, misguided doctor. William Blake understood that industrialised London, a city of darkness inhabited by miserable, sickly natives and rootless migrants who had left their pastoral idylls for a life of hardship in overcrowded slums, was enough to drive anyone to madness. In Catharine Arnold's elegantly written and richly anecdotal study, it is salutary to learn that it was not until 1890 that the status of lunatics was changed, by Parliamentary legislation, from prisoners to patients. Asylums were prisons disguised as hospitals, where the poor and incurable could be swept out of sight. It was a far cry from the charitable intentions of Simon FitzMary, who founded Bethlem in Bishopsgate in 1370 as a priory offering asylum to London's mad paupers. During the Crusades, he had been led to safety by the star over Bethlehem: the motif appears on the hospital's crest to this day. Traditionally, the medieval Church equated health and madness with good and evil. The mad were possessed by evil spirits, which could be driven out by beating, immersion in freezing water and periods in isolation. Sir Thomas More was as much in favour of thrashing the insane to bring them to their senses as he was of flogging heretics. Bedlam was racked by scandals. One inmate died after his intestines burst, having been chained in a confined space for years. Others slept naked on straw in the cold, tormented by sadistic keepers. There was money to be made out of the misery, hence the rise of the private madhouse. As the materialistic Victorian era gathered pace, Bedlam pushed its pauper inmates into new county asylums, making room for a burgeoning market of shabby, genteel inmates, driven to insanity by the pressures of middle-class life. Private madhouses were convenient dumping grounds for unwanted wives. Defoe noted that if they were not mad when they arrived, they certainly ended up so. By ancient tradition, the possession of a womb predisposed a person to insanity. Virgins and menopausal women were particularly vulnerable. One Victorian doctor advocated applying leeches to the labia, while another maintained that removing the clitoris saved a woman from insanity. It was no wonder, then, that the medics were perplexed when 80,000 ostensibly fit and active men suffered mental breakdown during World War I. They were not women, so why the hysteria? It was a further blow to conventional belief that most of the victims were officers, the elite drawn from the public schools. Accused of malingering, they were subjected to a new, barbarous electric shock treatment, before a more enlightened approach emerged. This is a thought-provoking book on a melancholy subject, with many parallels to the present. The mentally ill, like the poor, are always with us. The closure of asylums in the 1980s in favour of 'care in the community' proved disastrous. Many took to the streets, sleeping rough, as riddled with lice and despair as the medieval Bedlam beggar; others killed innocent citizens. In a final irony, some MPs want to overturn the Elizabethan legislation which bars those who have suffered mental illness from the House. How Swift would have loved that. To try to measure the hospital against others mental wards a comparison has been made.

Moorfields

The madhouse became so squalid and dilapidated by the middle of the 17th Century that in 1673 it was decided to move the hospital to a new, modern building in Moorfields - what is now Finsbury Circus. Finished in 1675, it was built into the north face of the City Wall and was generally thought to be one of the finest buildings in London. It was designed by Robert Hooke[3] and was the first purpose built hospital for the insane in the UK. Bald-headed and half-naked figures decorated the entrance gates, created by the sculptor Caius Cibber, and depicted 'Raving and Melancholy madness'. This is not the name of the statues; the title reflected the two categories of inmates. They reflect the names extremely well. 'Melancholy' looks defeated and vacant, and 'Raving' is chained to his plinth, a tortured expression on his face. The two figures now live in the Royal Bethlem Hospital museum[4]. A cast of 'Melancholy' is on view at the Museum of London. Made of Portland stone, the figures and hospital were one of the 'must see' sights of London at the time. It proved popular with those writing guidebooks and poetry, and engravers. The practice of charging entrance fee started around the time that the hospital moved, and a constant stream of visitors came to watch the patients. Not only locals but also foreign travellers and writers came to see the mad confined here. It was considered very important to the authorities that madness was seen to be managed and restrained, although many of the visitors who had not come to visit relatives ended up provoking the inmates for their own entertainment. All the visitors were charged a penny for the privilege. This was stopped in 1770 because it 'tended to disturb the tranquillity of the patients' by 'making sport and diversion of the miserable inhabitants'. Entrance was then by ticket only, designed to stop indiscriminate visiting. There were two galleries, each made up of a corridor lined with cells on either side. An iron gate in the middle divided the men from the women. The patients were often manacled to the floor. It was clearly a prison, rather than a hospital concentrating on curing the insane that it housed. Medical treatment at this time was largely ineffectual, and at this time patients were discharged after 12 months whether they were cured or not. Surprisingly, some inmates did recover. They were not referred to as 'patients' until 1700. As the hospital population grew, two extra wings were built between 1725-34. These became the 'incurable' wards, one for each gender. Once they were open, those patients who were released uncured were placed on the 'incurable' list. Once a place became available they were readmitted if they had no-one else to provide care. Patients were never admitted directly to the incurable wards, they always had to spend 12 months in the main section.

St George's Fields

By the end of the 17th Century the new Bethlem hospital was as decayed and desolate as the original had been. In 1807 it was decided to move the hospital again, after the building had been ruled unsafe - one wing had been demolished two years before for safety reasons. Having been built on the ditch outside the City Wall and on an area that had once been marshland, it had no foundations. The building work had been rushed and poor quality materials used. It was literally coming apart at the seams. It was moved to a new building at St George's Fields in Southwark. The marshy common land known as Saint George's fields spread north and south of what was then Lambeth Road and is now St George's Road. Moving into an area filled with prisons such as the Marshalsea[5] and the Clink, Bedlam seemed to have come home. Built between 1812-15, the new building was just as impressive as the old one. It opened with space for 200 patients, although only 122 moved in when it opened. It had the land to double the size of the hospital when more money became available. This was not until 1838, when work on extra wings for the criminally insane began. Other additions included a great dome to top off the existing portico that was decorated with Ionic columns. Grand on the outside, miserable on the inside, the interior still reflected the conditions of the old hospital. The sculptures that had decorated the old gate came with the inmates and were housed inside into the vestibule. Damaged and eroded by the weather, they were covered by a curtain. Restraints and physical punishment stayed the norm - one patient remained chained for 14 years. By this time other institutions for the mentally ill had been established; private asylums were much in demand to avoid the public provision at Bedlam. Fortunately the hospital had its own water supply, so diseases such as cholera and dysentery did not add to the misery of the patients. It was not until the mid-19th Century when the hospital came under regular government inspection that the treatment policy was changed. After two inquiries which were severely critical of the system, treatment rather than punishment began. Patients were given jobs and other things to occupy them, as well as medical treatment such as chloral hydrate (a sedative and sleep-inducing drug) and digitalis (treatment for heart conditions). Wards were furnished with more thought for comfort and keepers were gradually replaced by, or became, nurses. With these changes came the penniless middle-class lunatics; the common poor began to be cared for local asylums in their own counties. In 1864 the criminal patients were moved to Broadmoor, which was built to replace the cramped and prison-like criminal wings at Bedlam. They were then demolished.

Home-based or church-based care

Before asylums,the burden of keeping vulnerable individuals rested almost entirely on loved ones. 'Mad' people who could not be kept at home wandered free, begging for food and shelter, and often - like Shakespeare's Lear or Ophelia finding none. In Europe a few small Christian institutions dedicated to sheltering the insane emerged in the early Middle Ages. London's Bedlam was the most famous. It did not hold more than two dozen inmates until the 1620s. A growing market economy in the 1600s and 1700s saw 'service professions' emerge. Those who worked in them did thankless jobs formerly handled at home or by the church, and included undertakers, private tutors and 'madhouse - keepers'. Families paid for secrecy and discretion, and private 'madhouses' left few records. Artefacts show keepers used physical restraints such as leg-irons and manacles. Some keepers adopted 'management' techniques developed by Renaissance horse - masters to control stubborn horses. Quaker William Tuke founded the York Retreat in the 1790s. It was the first asylum to shun physical restraint and coercion. Its influential methods became known as moral treatment, which relied on constant surveillance. Around the same time physician Philippe Pinel famously unchained mental patients in Paris asylums, declaring they were sick, not criminals. This story is more legend than fact. Nevertheless, Pinel was a hero among asylum reformers and promoters in the golden age of asylums that followed. Public funding poured into asylum construction between 1800 and 1900. Advocates for building asylums included Dorothea Dix. Like many Victorians, they placed faith in bricks-and-mortar solutions to social problems associated with the accelerating pace of modern life. The patient population in England went from about 10,000 to 10 times that. Asylums built in this period impressed public officials, as the buildings were designed to be majestic and therapeutic. Extraordinary attention was paid to ventilation and to safety, and most asylums also featured extensive grounds beautiful gardens were tended by patients as part of their treatment. Psychiatrists combined Tuke's and Pinel's methods to manage patients. They created a medical version of moral treatment reliant on the secular moral authority of the (always-male) psychiatrist, who controlled asylum life. The theory was that a carefully designed and regulated environment, with a strong father figure, calmed patients and restored their sanity. Asylum physicians' exaggerated claims of curing 'lunacy' by moral treatment backfired. Expensive but inflexible buildings became overcrowded, and by 1890 the majority of patients left only in coffins. Old techniques returned - straitjackets, seclusion and sedative drugs such as bromides were used on unruly patients.

Controversies and legacies

In the first half of the 1900s asylums (or 'mental hospitals') became testing grounds for controversial treatments such as electroconvulsive therapy (ECT) and lobotomy. These methods helped some patients function again, but they irreparably harmed others. Such therapies became widely used because doctors and nurses wanted to offer patients cutting-edge treatment. ECT and lobotomy, however, reinforced an old and persistent image of asylums as intimidating places of last resort. Many mental hospitals closed in the 1970s and 1980s. This was due to pressure from the antipsychiatry movement, feminist criticism, ex-patient activism and political suspicion of large, unaccountable institutions. Other mental hospitals were converted to 'short-stay' treatment centres - a policy enabled by new psychiatric drugs. In the UK this was called 'care in the community'. Many patients were left homeless. Others, especially people with profound intellectual disability or brain damage, remain institutionalised in 'care homes' their entire lives. Such patients and ex-patients depend on loved ones or charity to weather political and economic changes. People without such shelter are often forgotten. This distantly echoes the situation of people called 'mad' in the Middle Ages. The hospital moved to its third site in 1815, south of the Thames to Lambeth. Part of the grandiose, enlarged establishment survives now as the Imperial War Museum. In 1930, Bethlem Royal Hospital settled into its current site at Beckenham, near Croydon.

Bedlam Bodies

Archaeologists have now discovered as many as 1,000 bodies at the hospital's former burial site near London's Liverpool Street station, as part of the capital's Cross rail development. While some of the remains will ultimately go on display in the Museum of London, Government regulations demand the bodies are ultimately reburied locally – though indecision still surrounds the exact location. "We are talking about several hundreds, possibly thousands of sets of remains," said Cross rail archaeologist Jake Carver. "We have made a larger hole at the site than anything previously created here."

The discovery dwarfs the recovery of 400 bodies at the site during the 1980s as part of the development of the Broadgate Centre. Many of those remains were reentered beneath the Centre itself, though Mr. Carver predicts the larger numbers of corpses this time around will make that impossible. "We need to agree a suitable place to rebury them," continued Mr. Carver. "In similar situations it has been the East London Cemetery which has had the space. But we will not finish the work for another year or two. Prior to the corpses' re-internment we will try to undertake analyses to find out more about those buried there. What genders were they, what ages, and did they suffer from particular pathologies?" The site lies beneath the location of Cross rail's future ticket hall at Liverpool Street. It was once occupied by a cemetery identified in historical records as the "Bethlehem Churchyard", part of Bethlehem Hospital's land holdings. "Bedlam" was an old abbreviation of "Bethlehem" and became associated with the institution, which for much of its history was known for its inhuman treatment of patients.

Nathaniel Lee

(c. 1653 – 6 May 1692) was an English dramatist. He was the son of Dr Richard Lee, a Presbyterian clergyman who was rector of Hatfield and held many preferments under the Commonwealth. He was chaplain to George Monck, afterwards Duke of Albemarle, but after the Restoration he conformed to the Church of England, and withdrew his approval for Charles I's execution. Lee was educated at Charterhouse School, and at Trinity College, Cambridge, taking his B.A. degree in 1668.[1] Coming to London, perhaps under the patronage of George Villiers, 2nd Duke of Buckingham, he tried to earn his living as an actor, but acute stage fright made this impossible. His earliest play, Nero, Emperor of Rome, was acted in 1675 at Drury Lane. Two tragedies written in rhymed heroic couplets, in imitation of John Dryden, followed in 1676, Sophonisba, or Hannibal's Overthrow and Gloriana, or the Court of Augustus

Caesar. Both are extravagant in design and treatment. Lee's reputation was made in 1677 with a blank verse tragedy, The Rival Queens, or the Death of Alexander the Great. The play, which deals with the jealousy of Alexander's first wife, Roxana, for his second wife, Statira, was a favourite on the English stage right up to the days of Edmund Kean. Mithridates, King of Pontus (acted 1678), Theodosius, or the Force of Love (acted 1680), Caesar Borgia (acted 1680), an imitation of the worst blood and thunder Elizabethan tragedies: Lucius Junius Brutus, Father of His Country (acted 1681), and Constantine the Great (acted 1684) followed. The Princess of Cleve (1681) is a gross adaptation of Madame de La Fayette's exquisite novel of that name. The Massacre of Paris (published 1690) was written about this time. Lee had given offence at court by his Brutus, which had been suppressed after its third representation for some lines on Tarquin's character that were taken to be a reflection on King Charles II. He therefore joined Dryden, who had already admitted him as a collaborator in an adaptation of Oedipus, in The Duke of Guise (1683), a play which directly advocated the Tory point of view. In it part of the Massacre of Paris was incorporated. Lee was now thirty, and had already achieved a considerable reputation. He had lived in the dissipated society of John Wilmot, Earl of Rochester and his associates, and imitated their excesses. As he grew more disreputable, his patrons neglected him, and by 1684 his mind was allegedly completely unhinged. He spent five years in the notorious Bethlehem Hospital. He lamented his situation with the following missive: "They called me mad, and I called them mad, and damn them, they outvoted me".[2] He recovered and was released only to die in a drunken fit in 1692. He was buried on the 6 May in St. Clement Danes, Strand.

Moorfields Bedlam

Until the mid-eighteenth century, the Moorfields Bedlam, designed by Robert Hooke, was the only significant example of a purpose-built lunatic hospital in Britain. Major features of Hooke's Bethlem, in terms of the setting, accommodation and treatment, provided the model used for other charitable lunatic hospitals founded in the eighteenth century and even the publicly funded county asylums in the nineteenth century. As such its estate was very influential in asylum construction, including the principal elements of the estate: the building, airing courts and forecourt. Robert Hooke was a close colleague of Christopher Wren and designed several other institutional buildings in London including the Bridewell Hospital (1671-78), also for the City of London; the Haberdashers Aske's Hospital, an almshouse at Hoxton (c.1690-93); with John Oliver, Christ's Hospital Writing School (1675-6); as well as several town and country houses. Hooke's Hoxton building was of similar design to Bethlem: a long, single-pile building with an elaborate central block connected by flanking wings to two pavilions. A large, grassed forecourt appears to have been used by the inmates for recreation and exercise, and was divided from the road beyond by a wall and central gateway, the whole layout in similar formal style to that at Bethlem. It is illustrated in Strype's edition of John Stowe's Survey of the Cities of London and Westminster (1720). Hooke only used the single-pile design for these two hospitals, not for his houses. The new site selected, close by Bishopsgate, at the head of Moorfields, was chosen for its "health and aire", the benefit of an ample, unsullied fresh air supply and its effective circulation being regarded by the Governors as the key to healthy surroundings. The poem Bethlehem's Beauty (1676) emphasised the perceived virtues of the new site's healing air:

The Approaching Air, in every gentle Breeze,
Is Fan'd and Winnow'd through the neighbouring Trees,
And comes so Pure, the Spirits to Refine,
As if th' wise Governours had a Designe
That should alone, without Physick Restore
Those whom Gross Vapours discompos'd before

The Governors employed the prominent architect Robert Hooke (1635-1703), who was actively involved in the rebuilding of London after the Great Fire, to design the building. Largely constructed by 1676, it was probably only the third purpose-built asylum, after one in Valencia (1409, destroyed 1512) and the Dolhuys in Amsterdam (1562). Andrews states, in connection with the intentions of the Governors, that they were "much more concerned with the 'Grace and Ornament of the Building' than with the patients' exercise or any other therapeutic purpose New Bethlem was constructed pre-eminently as fund-raising rhetoric, to attract the patronage and admiration of the elite, rather than for its present and future inmates, whose interests took a poor second place'" The. Like Bishpsgate, it was outside the north boundary of the City, although only just so. The site for the building ran parallel to the ancient London Wall, and only nine feet (3m) to the north of it, occupying open ground on the site of the old City ditch which had been filled in. The site formed the south boundary of, and overlooked, Moorfields, a series of substantial formal public open spaces laid out from 1605, which, although largely surrounded by development, formed a finger of open space which led directly out to the open fields to the north. Moorfields Bethlem was palatial in scale, even in terms of new constructions put up as part of the building campaign after the Great Fire, being intended to accommodate 120 patients. The about 540 feet (166m) long entrance facade on the north front was depicted by Robert White in an engraving of 1677 (external link to picture), shortly after construction, together with parts of the grounds. The single-pile building was of two storey over a basement, and showed Dutch and French influences in its elaborate external decoration. The patients were segregated indoors, at first with males on the ground floor and females on the first floor. The cells, for individual patients, led off galleries which served for communication and for exercise in inclement weather. John Evelyn was one of the many admirers of new Bethlem, describing it as "magnificently built, & most sweetly placed in Morefields". There must surely have been a service entrance on the south side of the building, between it and the City Wall, although the space between the two was only nine feet (3m). The grounds were divided into a large rectangular forecourt in front of the building, flanked by two smaller exercise yards. The whole was approached via the formally laid out and enclosed lawns of Moorfields, a fashionable recreational space for the local inhabitants which had been one of the first such formally designated public open spaces. Security at Bethlem was of great concern, as patients were perceived to be continually likely to abscond as the opportunity arose. As reliable staff to supervise patients were difficult to find, the Governors had to rely on making the environment itself provide the means for ensuring confinement. The first three reports on the construction of the building by the hospital's Committee of Governors were largely taken up with matters concerning the boundary wall that was to surround the hospital and its grounds, and to confine the patients. The existing London Wall was used to form part of the secure 680 feet (c.207m) long south boundary wall. On the other sides a wall was to be constructed at 14 feet (4.2m) high along the sections which bounded the airing courts, with a coping expressly intended to stop the lunatics escaping. The exception was the front, north, wall of the forecourt which ran parallel to the whole length of the building and divided it from the adjacent Moorfields. This c.420 feet (c.128m) long central section of the whole north wall would be only eight feet (2.5m) high, so "that the Grace and Ornament of the said intended Building may better appear towards Morefeilds" The lowering of the forecourt wall did not affect security, for the patients were forbidden to exercise in the forecourt. (Bethlem Royal Hospital Archives, Bridewell and Bethlem Court of Governors Minutes, 23.10.1674) The wall was broken by six evenly spaced panels of iron railings, each forming a ten-foot (c.3m) wide clairvoie intended to enhance the views of the building from the adjacent and impressively laid out Moorfields open recreational space. The views were clearly intended to impress the users of Moorfields, both nearby residents and visitors alike, and the visitors to Bethlem itself upon their approach. Clairvoie (clear view): a gate, fence or grille placed in an otherwise solid barrier to provide a clear view of outside scenery or, in this case, the building inside. The north side of the building and the forecourt are shown in detail on White's engraving , with a passer-by admiring the ensemble. The clairvoie panels were flanked by piers surmounted by stone pineapples. At the centre of the forecourt wall an elaborate triple gateway gave access from the formally fenced and tree-lined lawns of Moorfields to the north, between which the visitor approached. The portentous gateway was elevated above a flight of steps and surmounted by the life-sized statues of two figures depicting raving and melancholy madness, attributed to Caius Cibber. From here the visitor crossed the expensively paved and graveled forecourt to gain access to the main entrance at the centre of the building. There were numerous large windows in the north walls of the ward wings, flanking the central administrative block, allowing for the ample ingress of light and air. Those in the raised ground floor and first floor, in particular, gave the patients an elevated view of the forecourt, and beyond this of the designed open spaces of Moorfields. The boundary wall also enclosed the two exercise yards, which flanked the forecourt at the corners of the building. The forecourt was several times larger than either of the exercise yards, which were limited in their extent by, apart from the forecourt which divided them, the proximity of Moorfields to the north, the London Wall to the south, and development to the west and east. Two yards were provided "reserved for the use and benefit" of the inmates, one each for the separate sexes to exercise in. Those patients "well enough" were "permitted to walke the Yards there in the day tyme", so that they could "take the aire in order to [aid] their recovery". (Bethlem Royal Hospital Archives, Bridewell and Bethlem Court of Governors Minutes, 23.101674 and 5.5.1676, and Bethlem Committee report, 16.10.1674. Each yard was surrounded by the 14 feet (4.2m) high wall, topped with a "Coping _ intended to p'vent the Escape of Lunatickes". Both were laid out with grass and gravel plots of 120 feet (36.5m), with, set into the rear wall, a small pavilion with windows at first-floor level. (Bethlem Royal Hospital Archives, Bridewell and Bethlem Court of Governors Minutes, 23.10.1674 and White's engraving 1677. The upper level of the pavilions may have provided shelter for attendants supervising patients whilst allowing them an elevated view of their charges in the yard, with the lower level providing shelter for the patients. By 1786, Bethlem was noted for its "fine gardens" where the patients "enjoy fresh air and recreate themselves amongst trees, flowers and plants". Marie Sophie von La Roche, 1786

Tagebuch einer Reise durch Holland und England, being her diary. Sophie in London, 1786 A translation of the part about England was published, with an introductory essay by Clare Williams (the translator) by Jonathan Cape in 1933. Although there was no formal classification by symptoms, there was obviously a category of patients who were allowed to exercise outdoors. Those whose behavior was deemed to be too wayward or who were physically too unwell remained indoors. By 1740 the wings had been extended to west and east in L-shaped form, covering much of the site of the early airing courts. Provision for patient exercise was made by reducing the width of the forecourt, such that it only extended half way along each of the original wings. It had also lost the clairvoies formerly sited in the north boundary wall. The open ground formerly flanking the forecourt was given over to airing courts, surrounded by higher walls. The old gateway had been re-sited to the south, much closer to the front door and a curved carriage sweep open to Moorfields now provided more direct access to the entrance to the building, while dividing the forecourt into two compartments. This removed the need for visitors to cross the forecourt on foot to gain admittance. The opening up of the approach physically linked the main entrance to the building with the main axial walk of Lower Moorfields pleasure grounds. The Moorfields site was abandoned in 1815, when a new site was opened in St George's Fields, Southwark. The old building had for long been unsound, having been constructed very quickly of poor materials over the unstable in filled ground of the City ditch below the City wall. The building was demolished and Finsbury Circus was developed on its site.

Nothing more active was done until 1800, when James Lewis, the hospital surveyor, reported that the whole building was dangerous, not one floor being level, nor one wall upright .There were reasons other than the unsatisfactory state of the building which swayed the Governors to adopt their surveyor's suggestion to remove to a new site. During the last quarter of the 18th century a gradual improvement had taken place in the public attitude towards the victims of mental disease, a change due in part to the work of John Howard and other reformers, and in part perhaps to the mental attacks of the King. The inmates of

Bethlem were no longer looked on as a "rare show" and the whip was discredited as an instrument of treatment, but fetters and straw were still in use, and the Governors, though resenting external criticism and control (they spent £600 in opposing the Madhouse Bill of 1815 in Parliament), were aware that the administration of the hospital left much to be desired. The old building was too confined and its associations too strong to admit of any great improvements there, and the Governors decided to remove the hospital to larger premises where the patients could be separated into categories and given better conditions and where there was sufficient space to allow them outdoor exercise. From 1801 to 1807 various sites were discussed, and for a time it seemed probable that the remove might be to Islington, but in June, 1807, the President and Treasurer reported that they had viewed "certain land lying in St. George's Fields being the site lately occupied by the Dog & Duck and at present held by Mr. Hedger under the Bridge House Estates" and recommended it as being fit and proper for the purpose they had in mind. (ref. 218) The negotiations took another three years, but in 1810 an agreement was signed whereby, in exchange for the land in Moorfields, the City Corporation granted 11¾ acres of ground in St. George's Fields for the erection of a new hospital for a term of years equal to that still remaining on the Moorfields property (963 years) and at the token rent of a shilling a year. The ground, roughly triangular in shape, had some houses along the road frontages a few of which were demolished for the new building, but the remainder, together with any land not required by the institution, were included in the grant in order to provide a source of revenue for the hospital.

In July, 1810, the newspapers carried an advertisement offering premiums of £200, £100, and £50 for the three best designs for a new Bethlem Hospital. James Lewis, the hospital surveyor, George Dance the younger, and S. P. Cockerell acted as adjudicators and they awarded first place to the design submitted by John Gandy Gandy's notes state that owing to the swamp ness of the ground he did not think it possible to sink the building more than 3 feet from the surface. He proposed a basement storey for the domestic offices and 3 storey above for the patients, each storey to contain 108 cells, eleven foot by seven foot six, with day rooms and keepers' rooms. The food was all to be carried up from the basement by back staircases. He suggested that a pediment supported by six Doric columns should form a central feature; otherwise the building was to be plain brick with a stucco cornice, since he deemed any fine or decorative architecture to be out of place. His proposed elevation, preserved among the archives at Bridewell, presents a dreary expanse of brickwork

From: 'Bethlem Hospital (Imperial War Museum)', Survey of London: volume 25: St George's Fields (The parishes of St. George the Martyr Southwark and St. Mary Newington) (1955), pp. 76-80. URL: http://www.british-history.ac.uk/report.aspx?compid=65447 Date accessed: 18 May 2012.

The Governors sought the advice of John Bacon, the sculptor, as to the disposition of Caius Cibber's statues of Raving Madness and Melancholy from the gates of the old hospital which Gandy had proposed to place above the pediment. Bacon advised that apart from renewing the toe of one of the figures and cleaning off the paint no attempt should be made to restore them. In his own words, no "intrusive chisel of any modern Sculpture [sic]" should be "suffered to invade the surface of these specimens of Original Art." The statues were accordingly placed in the entrance hall where they remained until the 1850's. They are now in the Guildhall Museum (Plate 41).

In August, 1815, 122 patients were conveyed in hackney coaches from Moorfields to their new quarters. They must have suffered acute discomfort during their first winter; the system of warming by steam was installed only in the basement storey and the windows in the upper storey were not glazed so that the sleeping cells were either exposed to the full blast of cold air or were completely darkened. The deputy surveyor replied to the complaints of the members of the Select Commission that the omission of glass was deliberate, since it allowed the cells to be ventilated to obviate "the disagreeable effluvia's peculiar to all madhouses." The windows were, however, glazed during the summer of 1816.

While the plans for the building were in preparation, the Governors received a request from the government to provide accommodation for criminal lunatics who until then had been confined with ordinary prisoners in goal. The request was the result of a Select Committee report of 1807 pointing out the disadvantages of mixing lunatics indiscriminately with other prisoners. The expense of the criminal blocks, completed in 1816, was defrayed by the government, who also paid for the maintenance of the inmates. Accommodation was provided for 15 women and 45 men. The eastern end of the hospital grounds, abutting on Ely Place (now Geraldine Street) and the gardens of the houses in West Square remained open until 1828, when the Governors leased it for 61 years to the Governors of the sister institution, Bridewell, for the erection of "a House of Occupations for the employment and relief of destitute of both sexes." These premises, which were afterwards known as King Edward's Schools, remained in use until 1931, when the children were removed to Witley. The old schools were pulled down soon after.

The first big alteration to the hospital was made in 1835, when the male criminal block was enlarged to take another 30 men. By this time the ordinary wards of the hospital had become congested, and Sydney Smirke was engaged to plan an enlargement of the whole institution. Besides providing accommodation for nearly double the original number of patients, Smirke was asked to provide workshops for the male patients and laundries for the employment of the female patients, while retaining the symmetry of the front elevation. The plan on Plate 39 shows how he solved the problem by building wing blocks at either end of the frontage and two long galleried blocks across the garden at the rear. Workshops and storage sheds were erected in the front yards, but the administrative block in the centre of the building was left with an unimpeded view of the gardens in front of the hospital which had been enlarged by the diversion northwards of Lambeth Road (see Plate 39), and which Smirke laid out afresh. Smirke also designed the single-storey lodge fronting Lambeth Road.

The Home Office, under Secretary Sir George Grey, decided in the late 1850s to identify a piece of land on which to build a dedicated special hospital. The site at Crowthorne, part of the Crown estate of Windsor Forest, was chosen for being reasonably isolated, yet also easily accessible from London. Crowthorne itself barely existed at the time, but Wellington College was being built nearby and was due to gain a station on the London and South East Railway, so the area was ripe for development. Broadmoor itself was to be perched high-up on a ridge within the forest, commanding a magnificent and suitably healthy view across the countryside below. Plans were shelved briefly when the Whig Government fell, and Grey removed from office, but as a result of Parliamentary enquiries into lunacy, it was not long before the Criminal Lunatic Asylums Act 1860 was passed. This allowed the Government to act on its plan and fund construction of its own asylum. Sir George Grey was back in post by the time building had begun, and under his instruction the Home Office's prison architect, Sir Joshua Jebb, was given the task of designing the structure. Within three years, an army of convicts had supplied their forced labour, the woods had been cleared, several brick boxes reached up to the sky, Jebb was on his death bed, and Broadmoor was open for business. For the first nine months of its existence, Broadmoor was a female only hospital. This was because the site design included fewer buildings on the female side, and they were finished first. The one female block was in a separate compound to the five male blocks that made up the initial building phase (a further block for each sex was finished within the next few years). It was only when these five blocks were ready, and the remaining convict labour retrenched to what would become Block 6 that coaches of men from Bethlem and Fisherton began replicating the women's arrival. That process began on 27th February 1864. Patients like Oxford and Dadd were amongst those transferred. By the end of 1864, there were two hundred men and one hundred women in the Asylum, though the numbers would swell further until there were around five hundred patients at any time, in a ratio of roughly four men to one woman. Of course, the social mix within the walls varied from month to month and year to year over the duration of the Victorian period. However, some statistics from the first year's intake of patients serve to give a flavor of the typical make up of this unique community, and how they had ended up there. Around a quarter of the men and 40% of the women were murderers; many others had attempted to kill. Otherwise, the average patient had probably been caught stealing, or, if he was male, setting fire to something. There were sex offenders too amongst the men, including pedophiles and those who had committed bestial acts. Those were all demonstrably serious offences. The law though, could pass a pleasure sentence for any crime, with the result that a relatively few patients were also treated at Her Majesty's Pleasure on what appear to be more trivial matters, such as vagrancy, sending threatening letters or even attempting suicide. Most of the men had been labourers or tradesmen in their previous lives, though around one in ten had served in the forces. The latter figure included soldiers and sailors who had seen actions in the campaigns of Empire across the globe. The professional class was represented too, including by patients such as Dadd, with his intellectual and artistic background. In contrast, most of the women were housewives or labourers, with comparatively few women coming from more privileged backgrounds. The suggestion has been made that, since many women had attacked their own children, the middle class Victorian lady was not likely to be found at Broadmoor. Any murderous tendencies such a lady might have had would have been deflected by her distant relationship with her offspring, and thus thwarted by the presence of the nanny. The poorer nature of the female class is perhaps reflected in the fact that while around two-thirds of the men were married, fewer than half the women were. These were workers more often than homemakers. There was a disparity in education too: most of the men could read and write, but only a third of the women could. Attempts were made to categories the patients, much as diagnoses might be made today. One of the tasks that befell the Victorian doctor in lunacy was to ascribe a 'cause of insanity' to each case. Sometimes these were what were termed as moral circumstances, such as: intemperance and vice; religious excitement; being unlucky in love; anxiety; and poverty. Yet even with the Victorians' fondness for morality, most causes were assigned to physical conditions, even if these were not fully understood, such as fever, head injuries and childbirth. Patients were also categorized by the activity that they undertook as part of their treatment. And, although the popular conception is that the Victorian asylum was a house of raving madmen, in reality around a third of the patients were well enough either to be employed in the Asylum or in its farm. If that serves to give a flavor of who was within the walls, it does not answer the question of how they came to be there. In keeping with the nature of Broadmoor, this question has both a legal and a medical side to the response.

Escape from Broadmoor

Escape from Broadmoor' is actually the title of a post-war British short, starring John Le Mesurier as the patient on the run. The film has nothing to do with the real-life Broadmoor, but the existence of the film title is good evidence of the fear that an escaped lunatic can cause to the wider community. This has been true of Broadmoor since it opened. Of course, whenever there is an element of coercion to keep people in one place, there will inevitably be some whose thoughts turn to being elsewhere. Victorian Broadmoor was not somewhere that most patients wished to make their home: it admitted them not by petition, but by the order of the courts or of the Home Secretary. So those domiciled in the Asylum were not necessarily willing guests, and most were sufficiently aware of their situation to object to it if they chose to do so. Some lunatics embraced this power more actively than others.

Victorian Broadmoor's record on escapes has to be seen in context. The relevant comparison at the time was to Her Majesty's prisons, or the county asylum network, and when this comparison was made, the new Criminal Lunatic Asylum had an enviable position. This was a fact that its Superintendents could parade before the Home Office when things did occasionally go wrong. The public perception of danger was always much greater than the reality, and eventually, Broadmoor's record was exceptional. In hindsight, however, this was a hard won reputation after an eventful first decade or so of the Asylum's life.

When Broadmoor opened in May 1863, everyone expected escapes to be attempted. Indeed, the site had been chosen so as to be reasonably close to London and the railways, but far enough away from other property that it would take an escaping lunatic some time to find civilization. Preparations for public protection were made onsite by barring the windows and erecting boundary walls. The staff lived mostly on the premises, and the patients were required to wear a uniform of grey clothing, marked on the lining with a crown and the Asylum's name. There were strict rules about what items patients could have access to, and handover systems were in place for staff so that no patient should ever have the opportunity to escape. There was also cure, as well as prevention: shortly after Broadmoor had opened, the Asylum wrote to the Home Office asking for authorization to pay reward money to anyone bringing back an escaped patient. The Home Office duly obliged, and suggested that they would be prepared to pay up to five pounds as a reward. Although these actions were essentially practical rather than strategic, they put on an impressive show of Victorian risk management in action, based on experiences in other custodial institutions. Whether every eventuality had been covered would only be tested by real-life attempts, and it was not long before the patients began finding the flaws in the system.

Over time, it would mostly be the men who tried to discharge themselves, so when the first escapee came from the female side it was a more novel event than might have been supposed when it happened. The date was Wednesday 8th June 1864, and it was late at night. Mary McBride woke up from her dormitory bed in the female block, went through an unlocked internal door into the ladies' chapel, jumped down from one of the chapel windows and ran off across the estate. It had been a remarkably straightforward escape. Not only had the dormitory door been left open, against regulations, but also there were no bars across the chapel windows, and once McBride was in the women's airing ground, she found only one wall, roughly six foot high, between her and the outside world. Such simplicity was not to be lauded, and action was taken promptly: the attendants in charge of the dormitory were reprimanded, and the absence of bars on the chapel windows was rectified within the month.

McBride was a fifty-one year old widow, a tall, thin woman with grey hair who had been convicted of theft at the Lancaster Sessions in 1857 and ended up in the county asylum there. She was a factory worker and, allegedly, a prostitute. Although notionally a convict patient, her sentence had expired five years before she had been transferred to Broadmoor. After her flight, her absence was not spotted immediately, and so a potential head start was afforded to her. She managed to make it as far as Reading before her apparel was spotted by a local bobby and she was retaken the next day. Broadmoor's Council of Supervision fined the two attendants ten shillings each, and paid two pounds to the Superintendent at Reading Police Station as a reward. McBride tried to escape again in November, when she was part of a walking party exercising in the wider grounds, and as a result she found her future movements restricted solely to the female block and airing court.

This first escape was typical of the opportunistic nature of many, particularly in the early years while systems were still being established. So when George Hage became the first male patient to flee, he was also able to take advantage of the half-built nature of some parts of the estate. Making off from the Terrace, at around seven o'clock on the evening of 19th September 1864, he had passed through a gate which had been left open from the Block airing court, and from there he went to the Asylum boundary wall, temporarily knocked down while the water tower was being built. Once again, it proved a simple exit, with Hage ambling out of the gate, away from the Terrace, and through the dismantled wall.

Hage was a young man, aged just twenty-two, with distinctive, auburn hair and hazel eyes, who had been convicted of theft at Leicester in 1861. In jail, he developed delusions that he was poisoned, so was removed first to Bethlem and then to Broadmoor. He lasted a little bit longer than McBride outside, working in a coal mine for a few weeks before his distinguishing features were recognised by Police in Sheffield, and he was re-admitted on 8th November. His escape, though, led to a minor scandal, when he confessed that an attendant called John Philport had agreed to turn a blind eye to his run. Philport was a prime example of the unreliability of some of the staff during Broadmoor's early years. Recently appointed, he had been trouble to the Asylum throughout his brief stay on its establishment. He had already been found to be so neglectful of his duties that only a week before Hage's escape; he was given notice to leave his post at the end of the month. The authorities' mistake was in allowing him to remain on site at all, and he carried on misbehaving to such an extent that he was eventually dismissed summarily before even his notice period had expired. In between these two disciplinary measures, and unknown to anyone, he had intentionally assisted Hage with the plan to allow the patient's liberty. After Hage had implicated the now ex-attendant, Medical Superintendent John Meyer turned the case over to the Police. They located his errant employee, arrested him and charged him, and in due course Philport was given twelve months' hard labour at the Reading Assizes. Hage, fresh from his gainful employment, was certified as sane, and sent off to Millbank Prison to serve out the rest of his sentence.

These first two escapes set the general pattern for future years, where a lone patient would first formulate, and then execute a plan which they hoped would lead to their freedom. Though most plans were of the moment, seizing on a chance to run, some tactics were thoroughly prepared. There was no noticeable benefit to either approach, as the level of preparation involved did not statistically make a difference to success. Yet while the lone lunatic runner was the norm, and it was exceptional for patients to conspire in concert, the third and last attempt of 1864 would also be the only one in the Victorian period that might be described as a 'mass breakout'.

Even then, it was only four patients who were involved: Timothy Grundy, Richard Elcombe, John Thompson and Thomas Douglas. It was Grundy who was the ringleader, 'a powerfully built man' according to William Orange, Meyer's then deputy. Accused of drowning his sweetheart after a quarrel, Grundy had been found 'not guilty by reason of insanity' at Worcester in 1863, aged twenty-seven. He was the first 'pleasure man' to try and escape, and had already been noted by the Broadmoor staff as a man who liked to try and organise direct action, being often secluded in his room for his troubles.

One Sunday in December 1864, these four men formed an elaborate plan. While the Chaplain was conducting an evening service on the ground floor for the men of Block 1, this little gang stood in the gallery upstairs, around the central staircase. They were not attending prayers. Rather, Thompson, a professed atheist, asked the attendant on duty if the latter might fetch a small piece of pie that Thompson had left in the ward. When the attendant obliged, one of the men shut the gallery door behind him and jammed the lock with a stone. The four of them then made their way into the first floor day room, barricaded the door, broke a window, and took out knotted ropes made from handkerchiefs, with which they proceeded to shin down the wall. Suddenly, it was raining lunatics. The Chaplain, part way through his lines below, looked up to see four burly figures passing by the ground floor windows. The alarm was raised, whereupon it was discovered that an accomplice, presumably on a given signal, had similarly stuffed stones into all the external door locks for the Block, effectively locking in the attendants and preventing a chase.

Fortunately, the stones merely delayed the staff from getting out, and the fleeing patients had not managed to exit the airing court before they were caught. It had been a near miss on this occasion. Block 1, together with the later Block 6, formed what were termed the 'back Blocks', for the more 'refractory', or violent patients. A breakout from one of these Blocks was likely to have greater potential consequences for the public. It was clear from the events of that Sunday that the back Block design required improvement. Meyer gave instructions to create an additional, more secure entrance to Block 1 from the administrative block, to ensure that one route out would never be blocked in the future; the design of Block 6, under construction at the time, was modified accordingly.

This was the first proper alteration made to the original specifications for the Asylum which had been brought about by an escape attempt, though more would follow with the successes of serial escape essayer, Richard Walker. Walker was probably one of the most difficult patients under Meyer's command, though this great scourge of custodians was not an exceptional man: he was five foot eight inches tall and of normal, if robust build. A thirty-six year old postman, who had stolen two letters in 1864, he had been sentenced to ten years in prison. Ending up in Millbank, he too believed that he was poisoned and in consequence of his delusions had recently arrived at Broadmoor.

In 1865, Walker tried his luck at egress a grand total of three times, on 8th April, 21st May and 3rd October. As might be concluded, he was unsuccessful on every occasion. On his first sortie, he and another patient, a Scotsman called Peter Waldie, managed to slip away from the attendants in Block 3 at twenty to eight in the evening. The only logical explanation at the time was that Walker had somehow managed to obtain a skeleton key, and then bided his time before taking his chance, but no key was found on him. The pair, still in their Asylum clothes, managed to walk as far as Bracknell, where they enjoyed a pub meal before being spotted the next day, lying down on the benches at Bracknell Station, by Broadmoor's gardener.

Walker was readmitted not to the comparatively genteel surroundings of Block 3, but to the back Block 1, in theory at least a more secure part of the Hospital. Not that Walker paid any heed to theory: he had developed a taste for freedom, and after lights out on 21st May he began to put a new plan into action, which was both detailed in its cunning and also not wholly thought through. This time it began with pebbles. Stuffing the lock to his door full of small stones that he had garnered from the Block's airing court; he then turned his bedstead on its end and placed it under his window. He stood on it, reached towards the high, small window in his room and broke the glass. Next, he passed his hand outside, whereupon he was able to unscrew the retaining nut and bolt of the centre circle of the window frame and bring them back inside. Using his new tool, he smashed the rest of the glass, until before him remained only a window-shaped hole. It was just large enough for him to squeeze his frame through. Once he was out, he dropped into a yard adjacent to the Block, and from there he could scale the six-foot boundary wall. He made for the Asylum stables, he found a horse, clambered up onto it, and rode off in the dark to Yateley.

So far, so good, yet in all this planning Walker had overlooked one small, but significant detail. Throughout his escape, he was wearing nothing except his nightshirt. When he duly arrived in the nearby village, it was half past four in the morning and he was naked from the waist down. Whichever way you looked at him, Walker must have stood out in at least one crucial area. What was a man on the run to do? He sought assistance. He came across a local carpenter, William Bunch, also up early in the morning, and told the tradesman that his unfortunate state could be explained by the occasion of his drinking with friends in London. Walker maintained that he was drunk, had missed his train and then been walking all night towards home.

Bunch took Walker round to the village postman, with the initial intention of getting his new acquaintance a lift to Blackwater Station. The three men sat in early morning half light in the postman's stables, where a jacket and trousers were found to cover Walker's modesty and some bread and cheese supplied for breakfast. However, their companion's appearance and behavior had immediately given both Bunch and the postman some cause for alarm, and they kept Walker talking while separately, a messenger was sent to the Asylum to check that no one was missing. A party of attendants then headed for Yateley, and Walker was back inside Block 1 in time for lunch.

Walker's last attempt of the year was also made with a comrade, Thomas Douglas, who was himself making his own second bid to abscond. This time, both men managed to gain access to one of the wards in Block 1, then broke through the window of a single room much like Walker had done previously in May. From there, they made out first into the Block 1 airing court, and then over its dividing wall into the airing court of Block 3. The alarm was raised at once, and the pair were found hiding in the coalhole of the admin block. It was clear by now that Walker owed his successes to more than just good planning. 'Walker has long been supposed to have had a key and this alone can account for his being enabled to pass through the doors', reported Meyer. He was quite right. Three months later, the key was finally discovered: an intricate piece of ironwork probably based on an impression made of a Broadmoor key by Walker or another and then worked up for the patient by a criminal associate outside. An attendant, suspected of helping Walker to hide the key, though not of being party to the escapes, was dismissed.

This third attempt was both the least effective, and the last of Walker's efforts, and it landed him in a form of solitary confinement for most of the next few years. Seclusion was the principal method of containing unruly patients, and now Walker found himself secluded as a matter of course. Though he was now safely secured, his management remained a great challenge, as almost uniquely amongst Broadmoor patients, the medical staff found Walker impossible to control. He was an insubordinate extrovert, and at times, he had an attack on sight policy. He would prowl around in Block 1, naked apart from a strip of cloth around one arm or leg, covering his room in faeces or using them as missiles with which to javelin the doctors when they visited. This made him into something of a cause célèbre for the Commissioners in Lunacy, as Walker was in consequence kept alone by Meyer in one end of the first floor gallery of Block 1, away from the other patients and with an attendant beside him at all times. The Commissioners lobbied Meyer to allow Walker greater freedom, believing this situation to be unpalatable, and a throwback to an earlier era of chains and restraint in Bedlam. So, depending on his behavior, Walker was sometimes permitted to exercise in the airing court, though always alone, where instead of company he had a collection of pigeons that he fed.

Walker's legacy was the decision to replace the cast iron bars in Block 1 with wrought iron bars and window shutters. This replacement would prove to be an effective deterrent, though the original bars were retained in the other Blocks for the time being, where the patients were felt to be less likely to attempt to destroy them. With the benefit of hindsight, that budget restriction would turn out to be a mistake, something that would be acknowledged only three years later.

These escapes had also highlighted a simple truth to the Broadmoor management. 'It is obvious that the walls dividing the different airing courts must be raised', wrote Dr Meyer. These walls were not the external boundary, but they did afford a patient the means to move from one part of the Asylum to another without impediment. Meyer was allotted the sum of £50 to raise all of them by three feet, so that their full height became between seven and eight feet, and some mechanical help would be required to climb them beyond a patient's own means. No further action was yet taken to raise the height of the boundary wall.

It was still felt that there was no need to do so. Any patient allowed outside an airing court was either lower risk, or being invigilated to such an extent that making it over the boundary wall was not an option. Any failings in this area were likely to be through human error. That hypothesis was strengthened in 1866, when Patrick Lyndon, a trusted patient, made his unsuccessful attempt at self-discharge.

Lyndon was the first pleasure man since Grundy to try and get away. He had always been keen to remove himself, whether by orthodox means or not. He regularly petitioned the Home Secretary for his discharge, and sought to place himself in situations where he might escape. He had been in hospital care for over twenty-eight years, had lived longer inside asylum walls than he had outside them, and was still keenly awaiting a word from Her Majesty.

In Lyndon's case, it was his motivations for Her Majesty's pleasure that had indirectly contributed to his present position. A native of Liverpool, Lyndon made the journey south to Buckingham Palace in 1838, where he presented himself as a divine messenger who had been instructed to marry the young Queen Victoria. It was not necessary to treat him as a king, he said, and he was taken at his word. Declaring that he had 'no earthly residence, not even an earthly name', he fought with the sentry on duty at the Palace Lodge and was charged with assault. He became a Bethlemite for seventeen years and was then moved onto Fisherton, where he was considered to be 'an industrious man', albeit one who had also escaped on more than one occasion there. Now, he was in his mid fifties. He was first put into the shoemaker's shop at Broadmoor, where he was not considered to be good at his work, and had been moved into the garden. It was the decision to place him in the garden that led to his temptation.

Usually the staff were vigilant of Lyndon, as his auto-removal tendencies were well known. Now though, when sending Lyndon on his way to the Asylum garden from its kitchen, attendant Henry Franklin did not bother with the normal handover of his charge to another employee. Lyndon had never presented him directly with any trouble, and the attendant was relaxed about the oversight required of him. It was not long before Franklin realised that he had made a misjudgment: rather than saunter down the path to gather vegetables, as was intended, Lyndon upended a wheelbarrow at rest in the garden, stepped onto it and jumped astride the boundary wall. A supple youngster might have vaulted straight over the wall and made for the woods, but for Lyndon, making it over the wall had been exertion enough, and he was spotted progressing at low speed by another attendant at work in his own garden, and wrestled to the ground. Franklin was admonished, and Meyer pointed out to him that were similar circumstances to arise in the future, it would be quite clear where the blame would lie.

By the end of 1866, then, some remedial work had been undertaken to greater secure the Asylum, and better practice was beginning to result from the experience of staffing it. There was a brief respite in the frequency of patients trying to absent themselves from care. At first sight, the decrease in the rate of escapes implied that the security systems, particularly the buildings, had been shorn successfully of their original defects. But it was not so, and the events of 1868 would demonstrate to Meyer and his staff just how much work there was still to do.

By the autumn of 1868, it had been two years since Meyer and the Council of Supervision had spent any sums on making changes to the buildings as a result of escapes. The window bars in the non-refractory blocks, identified as a weakness in 1865, had been left as they were, persistently passed over on the grounds of cost. That the decision to retain the cast ironwork in the less secure blocks was a false economy was about to become clear.

On the evening of 4th November, James Bennett, a youth of eighteen, removed a cast iron cross bar from the window of the ground floor gallery in Block 3 and made his way over the still-deficient boundary wall. Bennett became the first patient since George Hage to get away for a considerable period of time, and, despite the obvious blame that could be attached to the window bars, his escape led to the resignation of the attendant who was on duty at the time. It was considered that Bennett should have been spotted in such a public part of the building, and the implication was that the attendant's vigilance had been found rather wanting.

Bennett himself had come to Broadmoor in March 1867 as a depressed and suicidal young man. He had an unenviable start in life: suffering from mild learning disabilities, and evidently prone to anti-social behavior, he had spent three years in a reformatory school between the ages of nine and twelve. The sharp shock did not work: subsequently given seven years for theft in London, he had been sent to Portland Prison. In the month before he ran from Broadmoor, he had been fighting intermittently with another patient on the ward. When he escaped from the Asylum, he quickly returned to his old stamping ground in Chelsea.

From Bonfire Night onwards, Bennett had a full three months of freedom from Broadmoor before he managed to get himself arrested again, this time caught exiting someone else's property with a quantity of linen. Although he gave his name as 'William Watson', he also, rather honestly, owned up to the fact that he was wanted back in Crowthorne. The Westminster Police Court officials sent a message to Broadmoor and asked someone to attend court to identify him, which they did. He was returned to Broadmoor on 10th February 1869. Meyer subsequently concluded that also like Hage, Bennett had only been faking his insanity, and so he had his patient removed to Millbank Prison, whereupon Bennett's involvement with Broadmoor was over.

This was merely the start of things. Just as Meyer was beginning to fear that Bennett had been lost forever, another two patients disappeared from his radar. On 9th November, Thomas Douglas and John Thompson, survivors of the 1864 gang, broke a similar iron cross bar to that of Bennett in a single room in Block 4. That the two men had managed to secret themselves in a single room was due to timing. The Block's patients had just had tea, and the attendants were engaged in tidying away the crockery and cutlery. Douglas's and Thompson's escape was slightly more complicated than Bennett's in that they were on the first floor, but using their previous experience of escapes, they had ripped up the bedding in their rooms and then tied the pieces together to form a rope. Throwing it out of the broken window, they both shimmied down into the yard, and then up another blanket rope that they had dropped previously from a room nearby. This second rope brought them close enough to the top of the external wall that they could swing over and onto it. Once on the wall, they were down the other side and away. For Douglas in particular, this must have felt like the completion of a long-held dream. For Meyer, it was a nightmare. That evening, he was facing the unprecedented loss of three patients within a week.

Douglas, an ex-soldier, was a native of Cumberland who initially struck out south. He walked from south Berkshire to Southampton with the aim of securing a passage to America. He had been a sailor before he joined the Army, and still wore an anchor tattoo on his left arm as a testament to his earliest career. But sailor Douglas could not find a suitable ship at port, so he then decided to return to his native home in the far north. He walked the length of England, before eventually, exhausted and starving after nearly three weeks on the road, he gave up at Lancaster and surrendered himself to the Police on 30th November.

He returned to the Asylum a reformed man. Biddable and co-operative, he worked in the garden and asked to be returned to prison. His wish was granted in 1870, and he served the short remainder of his sentence for insubordination in Millbank. This was not, though, to be Douglas's last experience of Broadmoor. A little over a decade later, he was had up for assaulting a police officer in Portsmouth and given six months hard labour. Though he called himself Kelly, he was identified as Douglas and sent back to the Asylum, where William Orange, Meyer's successor, suggested he might be happier to remain. He spent the last twenty years of his life back in Broadmoor's care and died there in 1903 from heart disease.

Meyer's luck was in, and it continued to hold when Thompson was also arrested by the Police in Garstang, Lancashire on 7th January 1869. All three of these November deserters had come back before winter was out. It had been a close shave. It was also obvious now to those in charge that they had made the wrong choice about the windows back in 1865. After the escapes of Bennett, Douglas and Thompson, Meyer summed up the situation: 'The Council have long been aware that the cast iron bars and window frames which existed six years ago throughout the buildings were most insecure, and the evil has been remedied in Blocks 1 and 6 in which the windows have all been secured with wrought iron bars...there remain however 784 windows not yet secured...Mr. Jarvis, the clerk of works, estimates the expense at £1100 and believes that the work might be completed in two months'. He asked the Council of Supervision for permission to carry it out immediately. The Council agreed, but the price was a significant sum, and they had to ask the Home Office for the money. It was forthcoming in December 1868, and it came too late. This delay in rectifying a design fault was about to cause the Asylum what became, in retrospect, its first real embarrassment.

It was Christmas Eve, and patient David McLane was alone in his single room on the first floor of Block 4. Before him was his window, with its iron cross bar. To remove the latter, McLane used two pieces of metal from old locks, and a piece of wood to steady the pressure. Correctly applied, he had managed to turn the bolt in the window frame; taking additional advantage of the fact that one of the retaining screws in the frame was faulty. Once he was out of the window, nobody was sure exactly what route he took, but it seems probable that he managed to follow the roof line round the lower level of the administration block, reach the Gatehouse and then drop down outside. McLane had entered his dark chamber at seven o'clock on the night before Christmas, and was not missed until twenty-five to eight on the morning, by which time he was, presumably, long gone. There were no inspection windows in the doors of Block 4 at the time, and as far as the attendants were concerned, all the lunatics were sleeping peacefully on the first noel.

McLane was not the sort of patient to bring peace to all men. A violent offender, a rapist, he had been convicted at Durham in 1863 and sentenced to eight years in jail. After he had been moved from Wakefield Prison to Millbank, he had begun to hear voices in his cell, and also to believe that he was under the power of electricity, used upon him by forces unknown. Electricity was the new poison, for the development of technology is felt in delusions just as much as in the real world, and McLane was an early sufferer from the same discovered electric currents that would affect many other Victorian patients.

It seems that in the days leading up to his escape, McLane had been the fortunate beneficiary of a lapse in good practice: he had obtained clothing and boots without these being checked out to him properly, and had stored them in his room for when the time came. If the rules had been followed, then McLane would have escaped in only his nightshirt, like Richard Walker before. A half-naked man in the depths of winter may well have given himself up if he had been unable to find clothing outside. A fully clothed lunatic had already gained an important advantage. The Block's senior member of staff was severely reprimanded for his lack of oversight.

Delusional or not, McLane had evidently well-planned his escape: apart from the clothes, he had been spotted the previous two mornings removing himself early from breakfast to go and look out of his window – presumably to survey his route - but no relevance had been attached to his actions. McLane's fate remains a mystery: his sentence expired in the summer of 1871, and he was written off the Asylum books the following year. It was Meyer's failure, and he moved as quickly as he could to rectify it.

Though the horse had gone, the stable door was bolted when the ironworks on the windows in Blocks 2, 3, 4 and 5 were replaced in early 1869, making them as secure as those in the back Blocks. This removed one of the principal methods of escape from the Asylum entirely, and henceforth, any attempt to escape from inside a block would have to be considerably more complex. There remained, though, another small window of opportunity within the fabric of the complex, that of the Asylum's boundary wall, and it would be from here that Meyer had his second, and final loss when the only woman to be lost forever made her way out in July 1869.

Alice Kaye was not as fearsome a prospect as David McLane. She was a thirty year-old factory worker from Bolton, with a partner and three children, who had been existing close to the poverty line when she stole a pair of books and two gold rings. She was given seven years inside. Alleged to be feigning insanity, she was removed to Broadmoor in March 1868 on the grounds that she believed she was the Queen. She had been a peaceful patient, generally working hard in the laundry and on the ward, and not presenting many problems.

Her escape was entirely one of chance. At seven o'clock on the evening in question, Alice and roughly twenty-five other women were in the airing court of the new, additional female block. In the old block, the Asylum band was playing, and the female attendants in the new one were listening, some of them dancing with the patients. Sensing an opportunity, Kaye and another patient made their way towards the north boundary wall of the Asylum, which had been neglected while security had been improved on the male side. There had not been an attempted escape from the female wing since Mary McBride, and it was not felt likely that there would be another. Now, Kaye did what McBride had done. She got a leg up and a push and she was over the wall and away. Meanwhile, the band played on. She was only noticed missing when the final note had been blown, and it was time for everyone to go back in. But of course it was too late, and the vital minutes of song had given her ample opportunity to hide herself in rural east Berkshire.

Her description – brown hair, brown eyes, five foot one – was circulated to the Metropolitan and the Bolton police, and there was also a lead to follow up. Kaye had developed a close friendship with an attendant, who had briefly worked in Broadmoor a few months earlier, called Isabella Saby. Saby had, apparently, given Kaye an address in London and asked her to come and see her 'on the outside'. The implication of a relationship is hinted at in Kaye's Broadmoor notes. So Saby was tracked, visited and interviewed, but it was a blind alley: neither she, nor Kaye's family, provided information that they had seen the fugitive. Like McLane, Kaye was also written off the Asylum's books when her sentence expired.

Inevitably, more works followed Kaye's absenting of herself. The north boundary wall on the female side was raised later on that year, and the ground also lowered on the patients' side. John Meyer had finally achieved the basic levels of security that would have prevented most of the escape attempts he had endured so far. Though improvements were still required, never again would there be a lack of basic confidence in the accommodation provided to Her Majesty's lunatics. Unfortunately, Meyer himself would not have a chance to re-establish the Asylum's reputation for public safety. His sudden death, in

May 1870, brought to an end his time as Broadmoor's first chief of staff. With Orange promoted from deputy, the new Medical Superintendent immediately began agitating against the convict 'time' patients who he saw as the main source of disruptive behavior, including escapes. Statistically, he was correct: of the sixteen patients who had made serious attempts to escape under Meyer's tenure, only four were pleasure men. In this second phase of security development in the Victorian Asylum, then, the spotlight fell on another element: the lunatics themselves.

Dr William Orange, Broadmoor's second Medical Superintendent, had been a member of the staff since the Asylum opened. He had been part of the establishment that experienced the escape attempts of the early years, had directly witnessed some of them, and also knew the history of the protracted improvements to the window bars and external walls. On John Meyer's death in May 1870, he inherited an institution that had passed through an inevitable period of teething troubles in terms of managing difficult behavior.

Nevertheless, it would only be in 1875 that Orange finally felt confident that he had stemmed what was, admittedly, a gentle trickle of patients seeping out through the bricks and mortar. Until then, he would also suffer the indignity of reporting successful escapes to his superiors on the Asylum's Council of Supervision and in the Home Office. That Orange continued initially to fight against turbulence in flight was partly down to the building, again, but partly down to his more mature regime, with greater responsibilities and privileges placed upon both his patients and his staff.

During the early 1870s, though, Orange also strongly believed that many of his institutional ills could be attributed to the lack of segregation between his different classes of patient. Orange argued that the convict class of 'time' patient was far more destructive than Her Majesty's lunatics, the 'HMPs' or 'pleasure men' who were detained at Her Pleasure.

Both elements had been present from the beginning, and Orange now gathered his statistical evidence from Meyer's time to suggest that there might be some truth in the proposition that such patients were more prone to escape. This was part of a more substantive argument, for Orange also believed that the convicts' disruptive influence ran wider than this narrow problem. He felt strongly that the time patients were liable either to wreak havoc on their own in myriad ways, or to corrupt the mostly harmless HMPs. His hypothesis was affected by the fact that the numbers of both classes of lunatic had grown since the Asylum opened. By the time that Orange took over, the patient population at Broadmoor numbered over four hundred and fifty, which meant that his nursing staff of fewer than one hundred were significantly outnumbered by those who they were meant to watch. Around a third of these patients at any point were time sentenced, though the ratio of convicts was slightly higher on the male side. Orange concluded that as the numbers continued to grow, the potential for convicts to cause trouble would not diminish.

Though Orange felt that he had identified the building bricks of trouble, the potential escapees continued to come from both sides of the lunatics' dividing wall. Indeed, the first escapee with whom Orange had to deal was a pleasure man. It was on a frozen winter's day in January 1871, and a working party of seven patients and two attendants were labouring to break up the heavy soil in one of the fields on the Asylum estate, outside the walls. Isaac Finch, a thirty-one year old farm labourer from rural Essex, was a member of the group. Just before lunchtime, having finished his work and by now bitterly cold, Finch asked to be allowed to leave the party to return to his Block inside. He was given permission to cross a small bridge which divided the field from the enclosed part of the estate. Rather than make his way back through the gate, he seized his opportunity to run, and instead took off into the woods. The attendant in charge of the party was severely reprimanded. Orange was frustrated by his own inescapable position that higher security in the compound could always be circumvented by poor working practices.

Finch had spent most of his life as a member of the agriculturally disenfranchised, and had entered the Asylum as a married man with five children. His family life had been desperately poor, and the Finches lived only just above the poverty line. As he searched for hope and meaning in his struggle, Finch had become captivated by a form of evangelical Christianity preached by a group known as the Peculiar People. Their ministry was an Essex phenomenon, an offshoot from Wesleyan Methodism that promulgated a literal interpretation of the King James Bible, including the rejection of medicine in favour of prayer. The name of the sect was interpreted as 'chosen' rather than 'odd'.

A religiously conservative man, one summer day Finch had been found clutching his Bible 'with the leaves turned down at the death of Solomon and David', the son and father who, amongst other things, incurred divine displeasure through their sexual behavior. He was covered in blood. Shortly afterwards his wife's body was found at their home with her throat cut, presumably to relieve both her and him of their carnal sins. He was acquitted of murder on the grounds of insanity and arrived in Broadmoor in September 1870.

Now that Finch had escaped, Orange had this Christian murderer on the run, and it was only the patient's lack of organisation that spared his doctor's blushes. A pleasure man was, almost by definition, unaware of the consequences of his actions, and Finch's inability to act rationally was to be his undoing. Walking first to Windsor, then back westwards to Reading, Finch had turned once more and eventually decided to make for his old home in Essex. He tore off some of the Broadmoor labels from his clothing but did not complete the job, either forgetting to remove the rest or not identifying the need to do so. Reaching the Capital, and without food or shelter, an exhausted and hungry Finch had asked to be admitted to the Fulham Workhouse in Hammersmith, where his remaining Asylum markers were noticed by the staff. He was returned to Crowthorne only five days after he left, and the superintendent of the Workhouse's male ward was rewarded for his troubles.

Finch was not alone. Soon, Orange was dealing with another troublesome pleasure man, Thomas Cathie Wheeler. Born in 1824, he had broken down in his early twenties, after he had traveled to South America. He returned to England showing signs of profound character change. His family sent him first to Bethlem and then to the Surrey Asylum at Springfield as a voluntary patient, hoping that the respite would affect a cure. Discharged, seemingly recovered, one day in April 1852 he knocked his mother over with a flat iron, then took up a hatchet and beheaded her.

On the afternoon of 10th December 1872, Wheeler, now aged forty-eight, was amongst a group of patients from Block 4 who were strolling around the

Terrace to the south of the Asylum as part of their exercise routine. As it began to rain, the attendants in charge of the group started to marshal their troops back inside the Block, via its airing court. The patients massed at the gate, and filed past an attendant who was detailed to count the marchers as they went back in. His concentration, however, was broken when he noticed that a patient was attempting to smuggle in a stone inside a handkerchief, undoubtedly for use as a weapon at some later point. With the attendant focused on searching that patient, Wheeler acted on an impulse to conceal himself behind some large shrubs on the Terrace. He squatted down amongst the evergreens and waited in the wintry rain. Remarkably, he was not omitted from the initial head count when the gate was locked. Instead, he was able to wait until it was dark, whereupon he walked to a point where the boundary wall was lowest, found something to stand on, and climbed over.

Two hours later, Wheeler was eventually missed. It was enough of a delay to afford him a head start, and by now, he had begun to walk to the village of Blackwater, some three miles away. Unfortunately for Wheeler, the experience proved overwhelming. Frightened of losing himself in the pine woods along the route, he began to walk back towards the Asylum, intending to find and follow a different route away. Of course, moving back from whence he had come was associated with its own risks. As he approached Broadmoor, he was spotted by the Asylum's messenger, who managed to detain Wheeler in conversation for time enough until the duty attendants looked out of the Gatehouse and realised what was going on.

These cases were the exception, though, to Orange's convict rule. He was right to hold onto the evidence he had gathered, for he would be troubled much more by various efforts from his prison population to escape. Climbing onto covered walkways, running off over the cricket pitch, or the old habit of trying to break the windows; all these methods were employed by the convicts, and more frequently than any pleasure man. It was far more likely to find two convicts on the wall than one HMP in the bush.

Henry Leest's attempt was typical. He was a thirty year-old shoemaker from Pimlico who had been found guilty of theft in 1867, but was suffering from tertiary syphilis, which caused him gradually to become insane. In Broadmoor, he attempted suicide, endured hours of lonely seclusion due to his destructive nature, and attacked the Principal Attendant of his Block. Most disruptively of all, in April 1871 he beat upon Orange's new Deputy, Dr William Douglas so badly that the poor man was forced to resign through ill-health only six months after joining the staff.

So it was that Leest had already packed a lot into his time at the Asylum, when on 14th August 1871 he made off from a working party in the kitchen garden. Recently better behaved, he had spent the day digging up potatoes as part of a small group of patients helping with the harvest. Elsewhere in the garden, an attendant and another group of patients were shelling peas into baskets, while another attendant sat on a box nearby and kept a close eye on proceedings. Leest asked if he could go to the toilet, and was given permission to do so. Taking an empty basket with him, he made off towards the closets. The attendant watched him till he entered the building, seemingly thinking nothing of the basket's transport, and then turned his gaze back to the remaining workers. Leest was also keeping a close watch on things. When the attendant turned away, the patient used the opportunity to come back out immediately of the closets and to make his way to the edge of the kitchen garden. Placing his basket lengthways against the wall, Leest, a small man, was light enough for it to take his weight. He stood on the end of it and was high enough to grip the top of the bricks of the external wall. He was quickly over it and then away into the woods, leaving only the basket behind him as evidence. A pursuit followed within minutes, but came to nothing.

The Asylum had the address of Leest's brother in London, and they wrote to this gentleman to ask him for information. His brother was only too pleased to co-operate. The second Mr. Leest told the authorities that he had just received a letter from his escaped brother, and that it came with a Winchester postmark. Orange received this intelligence keenly, and at once supposed that Leest would make from Winchester for one of the southern ports. Attendants were dispatched to Southampton and Portsmouth to hunt down the fugitive. Orange was correct, and it was at Southampton docks that Leest was found, six days after his escape, waiting to board a ship to New York. He had managed to find work in the interim and had a week's wages on him.

It seemed quite clear to Orange that if Leest was employable and could operate a clear strategy for living, then he should be considered sane. Leest was sent back to Millbank as soon as the paperwork could be arranged. Eventually, Leest was even able to follow his American dream: a letter, probably written in the 1870s, survives on his file which was written to Orange from the distant shores of Cambridge, Massachusetts. Leest reported that he had been shuttling between Rhode Island and Boston on the Atlantic coast. Now he was writing to the Asylum to ask for money, because he was broke. No copy survives of the Asylum's reply, but if Orange obliged, and then it would not be the first time that such an informal grant had been made to one of his ex-parishioners.

It was patients such as Leest that convinced Orange that the convicts were a positive harm to his community of generally peaceful lunatics, their influence far outweighing their numbers. His first response was to take the most drastic action available to him, by increasing the number of patients, mostly convicts, who were forced to spend time isolated in seclusion. Many more hours were recorded of patients locked up in the day time, no longer able to roam. It was not in keeping with Orange's liberal regime, and it was not the solution. His harsh approach incurred the criticism of the Commissioners in Lunacy after their annual inspection, and he recognised that this policy did not suit him.

Orange changed tack. He reasoned that the only way to properly manage the pleasure and the time patients was to separate them entirely. His basic starting premise was that the pleasure men were innocents who had no wish to cause him trouble. None of them had ever been found guilty of a crime, and neither had society accused them of propensity to wickedness. It seemed only fair that they should be kept away from the taint of recognised offenders. Orange asked for new accommodation to be built, so that he could relieve the blameless. Delivering his annual report for 1872, he questioned 'whether it is just or expedient to permit those other inmates whose lives have not previously exposed them to such evil influences to be contaminated by the degraded habits and conversation of the convict class.' Over the coming years, he would try to achieve this separation where he could, but without any extra resources to do so.

While Orange battled his superiors, he experienced his annus horribilis in terms of escapes. It was as if all the charges he leveled at the convicts were about to be proved. It began with the worst experience of them all, on Saturday, 12th

July 1873, with the only murderer in Broadmoor's history to escape and to never be recaptured.

On that day, patient William Bisgrove was exercising in the Asylum grounds, accompanied by Attendant Allan Mason. For the patient, this was not a new activity. He had been allowed outside the walls many times, and he had been exercising in this fashion for about 18 months before. On this particular outing, Bisgrove and Mason strolled around the southern fields of the estate before turning, and making their way back towards the Asylum farm, pausing only to talk about the chickens that were running around their enclosure. As they moved on, Bisgrove pointed out some rabbit burrows adjacent to the footpath, and Mason, a big man, bent down to look at one of the burrows. Now that he was off guard, Bisgrove hit him hard on the back of the head with a stone in a sling, in the traditional patient manner. While the attendant reeled from the blow, Bisgrove attempted to throttle him, and then the two men grappled each other, before Bisgrove threw off his custodian and made his way, like previous runners, into the pine woods of Bracknell Forest.

Mason was temporarily incapacitated, but recovered and quickly made his way back to the farm. He raised the alarm, and then set off again in the direction Bisgrove had run. A thorough combing was made of the woods, but with no success. There were no leads until, as the searches were going on, word was received that someone fitting Bisgrove's description – a man with thick black curly hair and beard, and wearing the plain blue Asylum jacket and waistcoat with fustian trousers – had been spotted in the grounds of Sandhurst Military College. A search party spent the night there. Bisgrove was not found.

On the Sunday morning, a message reached the Asylum that Bisgrove had been seen in Aldershot on Saturday night. So throughout Sunday, a team of constables and attendants visited every lodging house and outbuilding in Aldershot, only to report back empty-handed once again. Then on Monday, a local woman told the Police that she had seen a man jump into the Basingstoke Canal two miles from Aldershot. The Canal was dredged, yet nothing was brought up that was connected to the fugitive. The Police back in Somerset were alerted, and the search closer to home was widened to Basingstoke, Winchester, Southampton and Portsmouth. Twelve days later, Orange called off the chase. It was probably sensible to stop wasting unnecessary effort: Bisgrove could be anywhere.

Bisgrove was quietly forgotten, though his description remained in circulation for a long time. Years later, in 1891, the Metropolitan Police asked Broadmoor whether they thought Bisgrove could be a man called James Sadler, who they had arrested for the murder of a Whitechapel prostitute (and who has occasionally been mentioned in connection with the Ripper murders), but the authorities were not convinced. It is an inconclusive end to the story, and Bisgrove's disappearance remains without a satisfactory coda.

Orange, a diligent and dedicated man, must have worried at the time that his errant charge was capable of committing a criminal act that would lead to their eventual reunion. For Bisgrove, an epileptic coal miner from Wells had a violent past. At the age of nineteen, he had spent his last free evening, a long August one, drinking with another youth and his girlfriend. Staggering towards home, they had reached a cornfield where they stopped. Bisgrove offered the girl two shillings if she would have sex with him, and she was inclined to accept. They laid down a short distance from an older man, George Cornish, who was asleep under the stars, and went about their business. As the other boy sat on a stile beside the byway, Bisgrove finished with the girl, got up, walked across the field and picked up a large and heavy stone. He carried it over to Cornish, slumbering sonorously in the summer night, and dropped the stone onto his head. Cornish died where he lay.

Bisgrove and his male friend were arrested and sentenced to death at the Somerset Assizes in December 1868. Both would have hanged, but Bisgrove confessed that he alone had committed the crime, though he had no recollection of it. His companion was set free and Bisgrove's own sentence was commuted to one of life imprisonment. The West Country adolescent had then been admitted as a convict to Broadmoor in early 1869, where he became one of that groups of patients who broke the windows and made physical threats to the staff. Despite this, during the last couple of years he had become calmer, hence his occasional strolls around the grounds, though as Orange noted, Bisgrove was 'always a morose and sullen man...inclined to recklessness partly from natural disposition and partly from there being so little apparently to be either hoped for or feared by him in this world'. A Victorian nihilist, Bisgrove's character was such that it seems incredible that he might have kept himself out of trouble for any great length of time after his escape, so perhaps this occasionally suicidal young man did end up at the bottom of the Basingstoke Canal in 1873 after all.

Less than a month after this unscheduled decrease in the lunatic population, and while Orange was away on a long weekend, the Asylum lost another patient. On this occasion it was the turn of John Walker, a thirty-five year old stonemason from Birmingham, to breach the staff's defenses. He had also been a difficult patient, with a long-term history of trouble. It might have all been very different: for when he was ten years old, he had taken his older brother's breakfast to the factory where the latter worked, then seen a mouse, chased it, and been struck on the head by the fly wheel of some industrial machine. Walker had suffered from learning disabilities ever since, and had been convicted of burglary in 1866 and given ten years inside. While in prison, he had begun to sense that he was controlled by witchcraft.

The circumstances of the case were similar to that of Bisgrove, in that Walker was being supervised outside the walls. On 7th August 1873, he was in a working party of eight patients in an oat field to the north west of the Asylum. The morning had passed without incident, and after lunch, the group returned to their labours. By four o'clock, the party had been at work for several hours, and they stopped for a break. The patients lined up, and the two attendants in charge poured out beakers of oatmeal and water for the men to drink. Walker was one of the first to receive his refreshment. By the time the attendants had reached the end of the line, they looked up to see Walker making his way towards the edge of the field. This was not unusual: they were some distance from the Asylum facilities, and if a man wished to spend a penny, the field edge was as good a place as any to do so.

As they watched, Walker reached the edge of the field, where he halted. Expecting to see him undo his trousers, their casual observation turned to alarm as Walker proceeded to vault the fence and, like so many before him, make off into the woods. One of the attendants immediately began to run after Walker, but caught his foot in a ploughed rut in the field and fell over. This gave the patient enough time to make good his sylvan flight.

It was a case of déjà vu. The usual searches were conducted of the woods and surrounding estates, the local police and the Met were informed, the railway stations were watched. He could be found nowhere. Walker was considered to be a low risk patient, and Orange suggested, perhaps sheepishly, that 'his liberation at no distant period would probably have taken place', but this was a further failing all the same. To lose one lunatic might be considered a misfortune, but to lose two most definitely had the whiff of carelessness about it.

Fortuitously, Walker did turn up again, though it was not until a little over five years later, and two years after his prison sentence had expired. It was a chance meeting between two old acquaintances. On 28th September 1878, one of the Broadmoor attendants was visiting Birmingham when he spotted Walker about the city. A personable conversation ensued, and the attendant suggested that it might be better for Walker to accompany him, in order to remove officially the cloud still hanging over his freedom. Even more fortuitously perhaps, Walker agreed, put up no resistance to returning to Crowthorne, and traveled back with the attendant the next day. Perhaps he felt that he had nothing to fear, as he had made a success of his time outside. After his escape, and as the summer of 1873 continued, he had taken seasonal work as a harvester, crossing England on a path from Berkshire to Liverpool. When winter arrived on Merseyside, he had gone back to his old job as a stonemason, moving back to his native Birmingham in 1874. At the time of his voluntary apprehension, he was earning two pounds per week and getting on well. It was quite apparent that Walker was sane and was also a productive member of society. It was in no-one's interest to stop his contribution. Orange discharged Walker absolutely three weeks later, gave him five shillings for his trouble and also paid his train fare back to Birmingham.

Orange could afford to be relaxed about escapees by the autumn of 1878. In 1873, the year still had another sting in its tail. It had its roots back in August too, when there had been a theft from the Principal Attendant's room in Block 1, one of the 'back Blocks' with higher security. Nearly fifteen pounds had been stolen, a large sum of money, and although searches had been made throughout the Asylum, the money had not been recovered. The reason was that a conspiracy was in progress. Despite a one pound reward on offer for information, the money was being hidden, quite possibly in turn, by two patients: Timothy Grundy and John Brown. Using the money they had thieved, both men also managed to bribe a corrupt attendant, William Phillips, into providing them with a skeleton key.

Brown was known as 'a very powerful man'. Stout, twenty-six years old, and serving a fifteen year sentence for wounding, he had attacked both staff and patients at Broadmoor since his admission in April 1871. He was another convict who had outgrown the cells at Millbank, and he did not find the regime at Broadmoor to his liking: 'I am weary of life in this cursed Bastille of misery and destruction', he wrote. He was often secluded in the Block, and for the two months before his escape he had embarked on a daily destruction of both fixtures and fittings on his ward. His behaviour improved in the few days before he escaped, almost certainly because now he had his key and wished to make use of it, and as a consequence he had been allowed around the Block again.

One day in November, he made out of the scullery on his ward in Block 1, opened the door to the Block with his key and went outside. He unlocked the airing court door, walked onto the Terrace, through another door and then into the yard where the wood was stored. He took two sets of steps, and placed a trestle over them. Then he climbed up, and over the wall, and made his way to Bagshot, where he spent the night in a cattle shed.

The staff were lucky. The next day, Brown used his money to make for an obvious destination that was being watched. He bought a ticket from Bagshot to Waterloo Station, and was retaken as he stepped onto the concourse in London. Giving a detailed, if varied account of his actions, it soon became apparent that Phillips was indirectly responsible for Brown's escape and the attendant was dismissed immediately. A little over five pounds of the stolen money was recovered from the inside of Brown's backside, and the patient was moved onto Millbank again the following year.

Eighteen seventy-three had been a bad year, the worst since Orange took over. Brown's case may have been dealt with, but the Medical Superintendent held an internal enquiry at the Asylum, directed by the Council of Supervision, to rake over the coals of Bisgrove and Walker, and to conclude with a report to the Home Office. Orange found little worthy of blame: the attendants in both cases might perhaps have been more vigilant, but in neither case were they negligent, and the principle that patients of good behavior should be allowed to go at large was not one that anyone who understood the subject wished to change. Furthermore, Orange did some research to show that the rate of escape at Broadmoor was around seven times less than for that body of criminal lunatics housed in county asylums.

Instead, Orange turned challenge into opportunity, and focused his recommendations on his lack of segregation for time and pleasure patients. He suggested implicitly (though none too subtly) that the matter of escape rates was related. The pleasure man's only chance of discharge rested on his good behavior, whereas the time man's reward for good behavior was to end up back in prison, which Orange considered to be a dubious incentive. He suggested that the pleasure patients be allowed to continue as they were, and that money be spent to provide a separate, more secure outdoors environment for the convicts. There was no mention of any changes to procedures that might have led his staff to improve their performance.

Orange's argument found support this time from the Commissioners in Lunacy, who agreed with Orange that his two classes of patient should be separated wherever possible. It was a victory, though not in the manner that Orange had intended. He did not get his extra money for extra buildings; rather, alternative accommodation was provided for the convicts at Woking Invalid Prison, Knaphill, and the next year saw an informal moratorium on time patients admitted to Broadmoor. Gradually, their numbers diminished, and, pleased with the improved level of control this gave him, Orange continued the moratorium, and trimmed the numbers further the following year. He also began purposefully to divide each Block into wards which contained convicts, and wards which did not. If nothing else, this efficiency does seem to have made it easier for him to deploy staff resources where they were more likely to be needed, and by 1876, he declared confidently that the management of the time patients was no longer a problem.

The presumption in favour of sending time patients to Woking Invalid Prison would continue until 1886, the year of Orange's retirement. Never again would he have to deal with as many admissions from the prison population. It was only when the decision was taken to close that prison that Broadmoor became the principal recipient of such patients once again. At this point, Orange's successor, David Nicolson, was given what Orange had asked for: funds to extend Blocks 2 and 5, and undertake sundry other improvements for better security, before the accumulation of convict lunatics in Woking made their way to Crowthorne in autumn 1888.

The effect of the separation was immediate, for there was only one attempted escape during 1874, right at the close of the year. It had cost money, time, personnel and the permanent loss of three patients to reach this stage, though the uncharted nature of the Broadmoor mission had surely ensured that such expenditure was almost inevitable. Yet before Orange could entirely relax, one further man decided to test his patience, and with a method of such intricate cunning that it should serve as a fitting finale to any story that ends with the sense that escape has become impossible.

A storm was raging around the Asylum on the night of 6th December 1874. The wind was swirling in between the blocks, buffeting the buildings on the forest ridge. The open gaps between the window frames and wooden doors were howling lowly with each forceful gust. In Block 6, if anyone had looked, they would have seen that patient Thomas Hart was busy worrying away at the wall of his room, inching ever closer to the other side. The catalyst for Hart's labours had been an unexpected discovery: that at the point where his bedroom wall abutted the bricks of a chimney flue, there was a much thinner skin of brick between him and the outside world, only nine inches thick instead of eighteen. Furthermore, after a little over a decade of weathering, the mortar joints had perished in parts of the outer course. Lying in his room, Hart could feel the freezing breeze, and he resolved to find out where it was coming from.

Hart was a destructive patient, and his bedstead had long ago been removed from his room. Instead, he slept on two mattresses, and it was this arrangement that afforded him the chance to begin to deconstruct the wall at its weakest point. Scraping away manually at the mortar, he managed to work first one brick loose, and then another. The noise of the gale ensured that no one heard Hart as he was working during the night. By placing his mattresses in front of his growing shaft, Hart could cover up his operations but also place the bricks that he removed between the two pieces of bedding. At the same time, throughout the night he listened out for the attendants, for by now each room had an observation hole in the door. He would be checked on roughly every hour; in between, he could execute his plan.

The patient was a twenty-two year old hawker, married with one child, convicted as a thief and serving seven years inside. At Broadmoor, he had been found to be impossible to employ at any ward or workshop task, but had taken instead to feeding the birds in the grounds of the Block 6 airing court, and to flying a kite for exercise. As a result of these recreational hobbies, Hart had been allowed to keep both a bag in his room containing bird food and a ball of twine for his kite wire. These items were about to be put to alternative use.

Hart worked throughout the night at his painstaking task. Then, shortly before six o'clock, he had pulled out enough bricks from the wall to create an aperture large enough for him to squeeze through. He gathered up all the pieces of discarded brick and put them into his bag of bird food. He dressed in a jacket and trousers that he had previously managed to secret in his room. He knotted his blankets together, and moved his bedding away from the hole. Then he took up the blanket rope, the twine and the bag and maneuvered himself through and down into the airing court.

If that was not ingenious enough, Hart's next moves were unsurpassed by previous escape attempts. Pawing away at the ground of the airing court, he scooped up earth and sand and added this to the contents of his bag, which by now contained a considerable amount of weight. He took a length of the kite twine and tied one end to the bag, and the other to his plaited blankets. Then he took hold of the other end of the blankets, picked up the bag, and swung the latter backwards and forwards until he had gathered enough momentum to throw it over the wall. It landed on the other side, still attached to the kite twine. This produced a cantilever effect, using the wall as the fulcrum. Hart had secured the heavy bag sufficiently to bear his own weight as he began to climb the boundary wall, gripping onto the blanket rope and easing himself one step at a time to the top.

He was missed at the hourly check at six, and search parties were immediately dispatched. Hart had begun to walk south, towards Blackwater, and he was spotted at half past nine in the morning begging for bread. A local labourer raised three friends, including the Asylum's coal man, and the four of them detained Hart on the road from Blackwater to Fleet. In the end, he was undone by one small element outside his control: the sole item of clothing that Hart had not been able to hide in his room were his shoes, and once chased, the barefoot patient was soon caught.

In consequence of Hart's escape, the Office of Works was instructed to examine the condition of all the other flues in the Asylum and rebuild them where necessary. During 1875, Orange also began the task of raising both the external and internal boundary walls, which divided the airing courts, to a height of between fourteen and fifteen feet, a protracted piece of work which continued into 1876. The inner compound was now over-engineered for safety, and the staff enjoyed a much higher level of confidence in the accommodation provided to Her Majesty's lunatics.

Hart's recapture can be seen as the end of the great escape period in the Asylum's history. Of course, it was not the end of escapes themselves: that day never came. The patients continued to make efforts to remove themselves, and always would do, but the successful conclusion of such plans became a rarer thing. In the remaining period of Orange's leadership, only one further patient managed to escape successfully, and even then he was recaptured the next day.

Since 1863, a total of eighteen patients had been able to help themselves to forbidden liberty, mostly for just a few hours, though with three evading re-admission in perpetuity. Most of these escapes had resulted in some direct alteration being made to the Asylum or the way it worked, and the level of public protection was increased continuously. Both Meyer and Orange learnt from the eighteen mishaps, and the result was, by 1875, a much more secure hospital. It was now twelve years old, more adult and fully-formed than when it had opened. Victorian Broadmoor was ready to receive greater numbers of patients, and to ensure that their discharge came about only via the due process of the law.

Broadmoor Babies

Broadmoor was no different to any other institution which housed women of childbearing age. Like a workhouse, a prison or a charitable refuge, it admitted women on a series of set criteria, regardless of their physical condition. The same was true equally of the county and city asylums which had sprung up during the nineteenth century, though with one notable difference. Although the average local asylum would have plenty of patients who had just experienced childbirth, those asylums very rarely received women who were pregnant, and who went on to have their babies within the institution. Generally, asylums were seen as somewhere to be avoided during pregnancy. Broadmoor, by comparison, showed its judicial side in these circumstances. As its patients had been deprived of choice in this matter in favour of direction, it had a small, if irregular number of confinements to manage. That these events were dealt with in-house, as just another part of ward life, was entirely in keeping with the ethos of the self-contained community that was Victorian Broadmoor.

To a large extent, the female side of the Asylum operated as an independent unit. The initial women's Block and its later companion were separated from the male side to the west by a high dividing wall. There was a dedicated body of staff of around twenty female attendants to nurse the residents of these blocks, with a female operational head, although the medical staff remained stubbornly male well beyond the Victorian period. The doctors' offices also remained on the male side, and in their charge, notionally at least, were the clinical interventions designed to remedy around one hundred lunatic women.

Male and female patients were barely aware of each other's existence. Work and entertainment were both provided separately. The result was that there was a parallel, segregated life going on for patients either side of Broadmoor's great divide. The women sewed and looked after the laundry, they promenaded along their Terrace or the wider grounds; they read in the day room and conversed; or, if they were in the female back Block, they were minded and managed as their aggressive counterparts were in the other half of the site. Even at the centerpiece annual events, such as the flower show or annual ball, the women were permitted only to mix with male staff, and not male patients. It provided both what was considered a safe environment for initial recovery, and also one where refuge could be given to help a patient to progress. It was into this single-sex regime that the women who arrived pregnant would find themselves.

The first patient to give birth in the Asylum was Catherine Dawson, who did so on 26th December 1866, a little more than two and half years after Broadmoor opened. That Boxing Day, at one o'clock in the morning, she was delivered of a baby boy in the infirmary ward in the female block. Her labour lasted only half an hour.

Catherine was in many ways a typical Broadmoor female patient. She was thirty-one years old, and a working class housewife from the industrial North West: Liverpool, in this case. The new baby was her fourth child. The older children had also been her victims. On 27th October 1864, she had cut the throat of her middle child, twenty-two month old Matilda, at the family's basic rooms in Toxteth Park, close to the Liverpool docks where her husband worked. She had also tried to kill her eldest daughter and had then attempted suicide. She was found insane before her trial, and given the pleasure sentence.

Although Broadmoor had opened eighteen months previously, Catherine was transferred initially from Kirkdale Prison to Rainhill Asylum (the Lancaster County Asylum) in Liverpool on 30th November 1864. It is unclear from her case notes why she was not immediately transferred to Broadmoor, as by that date the hospital had cleared its backlog of patients requiring admission from the older criminal lunatic accommodation at Bethlem, Fisherton House and other institutions. After Broadmoor's opening, it was unusual for a pleasure man or woman to be placed elsewhere, with incidents linked usually to the suffering of a temporary accommodation crisis; rarely, it might be because a patient was considered exceptionally harmless. Catherine evidently did not fit into the latter category, because she remained at Rainhill for fifteen months, until March 1866, when she managed to escape from the asylum. It took a month to track her down, though it was not difficult to find her. She was eventually discovered living once more with her husband, Henry, and the remaining two girls. She was brought back to Rainhill at once, and this time moved quickly to Broadmoor, on 15th May 1866.

On her arrival at Broadmoor she was instantly sick in the waiting room, and after her details were taken and her handover complete, she was confined to bed in the female infirmary, dosed up on beef tea and effervescing salts. The initial diagnosis was that her sickness had most likely been caused by a dose of morphine, administered to her to keep her calm on the train during the long journey south. That view held good for a few days, but when the sickness did not subside, the Broadmoor doctors concluded the true cause. During her month at large she had resumed entirely her marital duties, and had managed to become pregnant.

Catherine was probably not high up on the list of patients whose expectant condition would be easy to manage. She was an aggressive patient while she was in Broadmoor. She was quarrelsome and paranoid, imagining that tricks were being played deliberately on her. When her sickness eventually subsided she was moved to the Block's number two ward, then the ward for the more disturbed patients on the female side, and occupied her time with needlework and suspicion until she gave birth. The event itself was almost entirely unremarkable; in fact, the only remark Catherine made at the time of birth 'was that there was a nasty smell in the room'. Her baby boy was immediately removed from her after his birth and handed over to a dedicated attendant, who looked after him but reared him artificially on cow's milk. As Catherine was in no state to name her child, and the boy had been born on St Stephen's Day, the Broadmoor chaplain christened him Stephen. His baptism is recorded in the parish register for St Michael's, Sandhurst, presumably from a piece of paper supplied to the incumbent, as his mother's name was incorrectly recorded as Caroline.

The mother did not ask to see her child until a week after the birth, and it was not until two months had passed that she was finally allowed to see him. Their first, and almost certainly only meeting was not a success. Catherine behaved strangely with little Stephen, placing him on his legs to see if he would walk already and otherwise acting that he was older than a newborn, and the boy was taken away from her again on the same day, this time for good. It was clear that mother and child would never bond, but then it had never been intended that they should. As soon as Stephen had been born, Dr John Meyer, Broadmoor's first Medical Superintendent, had begun to plan arrangements for the baby's life away from his mother.

Meyer's plan was to ask either Catherine's local workhouse, or her husband Henry to take the child. He wrote to both. Henry Dawson replied most clearly: he was reluctant to accept his newborn son on the grounds of his own poverty.

Now lodging in Birkenhead, he was continuing to work while trying to feed the two surviving girls. He had a duty to the family that was in sight, not to that out of it. Meyer had more success with the Union. They had a series of questions for him as to their liability, but at no point did they refuse his request. After a little further correspondence, Broadmoor managed to persuade the officers of the Chorley Union Workhouse to take on the child of 'their' patient. A date for his removal was fixed, and Stephen was collected from Broadmoor on 25th February 1867 by the master and matron of the workhouse, and taken back to Merseyside.

The Dawson family was now split three ways. Catherine stayed in Broadmoor, her moods swinging between excitement and depression. When she was better, she kept in contact with her husband, reading his letters and writing her replies. But as well as her mental illness, she was often in poor physical health and unable either to write or to work at her sewing. She would lay in bed, exhausted, with her hands and wrists scarred from breaking windows in the female block. During one such period, in 1871, Henry worried that the long silence from his wife was fatal. He wrote directly to the Broadmoor authorities asking whether his wife were dead or alive. Shortly after it was confirmed that she was still alive, he visited her. He did not know when he would be able to do so again.

Though Catherine was slowly failing, it was Henry who died first, on 18th June 1872. A friend of the family wrote to Broadmoor to pass on the news, and Catherine was informed. Up in Birkenhead, the landlady of the house that Henry and his two surviving daughters had lodged in now took the remaining children on herself. Other friends took Henry's place as correspondent, but not to Catherine. Instead, they continued to write to Broadmoor, asking after the health of Mrs. Dawson.

Catherine spent the last two and half years of her life in the infirmary in the female wing, losing weight and becoming weaker. She was suffering from a degenerative disease. Her mind continued to sink with her body, and by January 1876 she had ceased speaking to the medical staff or being able to get out of bed.

There was one last moment of clarity. On 16th April 1876, she rallied briefly on her death bed. She spoke coherently, she chatted to her fellow patients around her. Then she died from tuberculosis, aged forty-one.

The story of the second child born at Broadmoor was a somewhat different one. Some fifteen months had passed since Catherine Dawson had given birth when Henry Meller arrived 18th March 1868. Henry's mother was called Mary Anne, and she was a stonemason's wife from Newington in South London. Mary Meller was twenty-seven years old; a small, stout woman with dark hair. She already had four children when she became pregnant once again in the summer of 1867. A few months later, on 1st November 1867 she attacked a widow who lodged with her and her family, hitting the woman over the head as she stooped to light the kitchen fire, and then trying to cut her throat as she sat down to recover. Her victim, Mrs. Mary Cattermole, managed to run from the house to safety, while Mrs. Meller tore at her lodger's hair and chased her into the street. Two men managed to tackle the assailant, and held onto Mrs. Meller until a police constable arrived to arrest her. Her trial in December was at the Old Bailey, and both her doctor and her father testified that she suffered from regular but intermittent bouts of insanity. She had attempted suicide on previous occasions. The prosecution made no attempt to press her guilt, and after a short hearing she was found not guilty by reason of insanity.

Despite the verdict, the Governor of Horsemonger Lane Gaol was not convinced. He wrote on her transfer document to Broadmoor that she was 'quiet and well-educated, betraying no symptoms of insanity'. Nevertheless he noted that she had attempted to poison herself while in his custody. She was admitted to the Asylum on 14th January 1868, seven months pregnant.

Mary was in better health than Catherine Dawson had been when her son was born. As a consequence, she was allowed to nurse her child for around three weeks before her husband, William, came to collect the baby and take him home. Mary was also noticeably improved since her admission, and though occasionally prone to physical outbursts, was employed regularly in needlework on the convalescent ward. Her change in character had been remarkable, and the Broadmoor staff suspected that it could be attributed to one thing: that she was sober. The possibility that it had been the drink that had driven her to attack Mrs. Cattermole had not surfaced at her trial, yet Mary was prepared to concede that it might be so. She confessed to previously intemperate habits, and even that she was drunk the night before the attack. Her experience was not uncommon to Victorian Broadmoor patients, several of whom had taken drinking to such a stage that the courts considered insanity to have intervened. In 1869, a report summarised her state as 'no doubt a bad-tempered woman but betrays at present no symptoms of insanity'. With a comfortable home and a caring, solvent husband, she was considered to be both well and at a low risk of reoffending. She was subsequently conditionally discharged into William's care on 3rd May 1870.

But this was not the last contact between the family and the hospital. In February 1873, William Meller wrote to one of the attendants saying that his wife had recently begun drinking heavily again. He complained that Mary was pawning the family possessions for money to fund her alcohol addiction. It was the letter of a man who felt that he had lost control of his spouse, detailing his inability to divert Mary from her errant behavior.

Amongst other tales, Mr. Meller recounted an evening when his wife had told the servants that she was going out to listen to a lecture. Since the venue was one where the couple had season tickets, with seats reserved for each event, Mr. Meller set off with the intention of joining his wife. Of course, when he reached the auditorium in question, both seats were empty, and Mary was not there. Distraught, William Meller set off for a nearby chemist's to buy some pills to calm his frayed nerves. As he waited for his tablets to be counted out, he chatted idly to the man behind the counter. The chemist mentioned that he had just seen a drunken woman pass his shop, pursued by a mob of 'a couple of hundred people'. Meller stopped dead: it couldn't be, could it? He raced out of the shop, following the direction in which the chemist had pointed, and shortly caught up with the mob. Sure enough, at the centre of the angry crowd he found his wife. Meller had no idea what she had been accused of doing, and was not particularly interested to investigate. He called a nearby policeman, who managed to disperse the throng, and Meller took his wife home in a hansom: 'but she would not sit in the seat and I was compelled to bid her lie in the bottom of the cab.'

William Meller asked Dr William Orange, Broadmoor's Superintendent, to write to his wife. He said that she took no notice of him, but he thought that she would take notice of Orange. About the same time, and apparently unconnected, Mary Meller wrote to Broadmoor herself. In it she asked Dr Orange to visit her. 'I am miserable and unhappy and require your assistance', she wrote. Her side of the story was different. She alleged that William had broken her nose, and stated that 'I would rather be under your care than be thus ill used'.

It seems likely that Dr Orange did write to the Mellers, possibly as a couple, as William addressed a further two letters to him directly in April 1873. It appears that husband and wife had managed to reach some kind of resolution themselves. Mary became more settled, and had been on a trip to Lancashire and Yorkshire. William also stated that Mary had brought little Henry home: whether or not he had been looked after by relatives up till then is unclear.

Although they had another child, the Mellers' family unit did not last significantly longer. Mary Meller would be another Broadmoor mother who died young. Her death occurred on 23rd December 1878 at the age of thirty-seven, and she was buried in Nunhead Cemetery in Southeast London. However, unlike Stephen Dawson, Henry had enjoyed an upbringing together with his parents and his siblings. He grew up to have his own family.

The Broadmoor staff had now experienced two quite different outcomes for the children born in their care. They would use these precedents to shape their future experiences. Their next chance to do so was three years away. This time, the mother was Margaret Crimmings, a twenty-six year-old single servant from London.

Unlike the other Broadmoor mothers in this story, she was a convict patient, rather than a 'pleasure woman'. She had not been found innocent by reason of insanity, but found guilty, and then developed mental health problems while in jail. Margaret had been sentenced to seven years' imprisonment on 11th October 1870 at the Middlesex Quarter Sessions. Her crime was stealing two coats, apparently from her brother. The length of her sentence was down to her past record, for this was not the first time that she had been inside. She had four previous convictions for theft on her file, the first at the age of eighteen, and a further one for assaulting a police officer. Already, she had spent a little more than two years of her life in prison.

The first few months of this latest and longest sentence were spent at both Westminster and Millbank Prisons in London. It was while she was here that the Prison authorities formed the view that she was insane, and asked the Home Office whether she could be transferred to Broadmoor. The matter of her pregnancy was an added complication, as it meant that should she move, accommodation would have to be found for her in the infirmary. Before the transfer was sanctioned, the Home Office took the step of writing to Broadmoor to ask directly whether the Asylum would be prepared to take her on.

Dr Orange replied positively, and she was admitted on 10th May 1871. This small, stout woman was eight months pregnant when she arrived inside the Gatehouse. Her skin was pale from her incarceration, and it contrasted with her dark brown hair. Immediately she was interviewed, and the Broadmoor staff unconvinced of her suitability for their care. Dr Orange wrote in her notes that she 'talks nonsense saying that she was frightened at Millbank and that I was the person who frightened her…it is evidently her desire to be thought insane at present'.

Nevertheless, she was here now, and was not about to be moved again. Her child was born soon after her arrival, at 5.15am on the morning of 8th June. The first girl to be born in Broadmoor, she was christened Margaret Julia by Broadmoor's visiting Catholic priest. Like Mary Meller, Margaret senior was allowed to nurse her baby at first, doing so 'in a sensible and affectionate manner'. But on 12th June something changed, and she began to act oddly, suggesting that she had known the attendants for many years, but that now they were using false names; that the nurse helping her was not holding the baby properly, intending to hurt it; and that people were being unkind and speaking badly of her. Diagnosed as having entered a maniacal state, her baby was quickly taken from her.

With no husband or partner to care for the illegitimate child, Broadmoor wrote to the St Marylebone Union, where Margaret had spent time in the workhouse during 1870, to confirm the guardians' duty to take the baby. They acknowledged their obligation, but reluctantly, and asked whether Broadmoor could allow the baby to stay with its mother until her removal back to prison. Dr Orange considered this to be of no benefit to the infant. He replied that 'the mental condition of Margaret Crimmings is such as to preclude the possibility of leaving the child under her care…as under any circumstances the child is deprived of its mother's care its removal from the Asylum would appear to be desirable on all accounts.'

So the Assistant Matron of St Marylebone Workhouse came to collect Margaret Julia on 19th July, and take her back to central London. Sadly, the baby girl was to have a very short life outside the asylum. She died at the workhouse nursery, Southall School, on 19th August 1871, when she was only ten weeks' old. The guardians wrote that her death was due to 'debility', an unspecific cause, though a description of Margaret Crimmings's teeth in her Broadmoor notes raises the possibility that both mother and child suffered from congenital syphilis.

Meanwhile, Margaret remained at Broadmoor, and was pronounced recovered from her mania by August. She was employed in the asylum laundry where she was an industrious worker, occasionally prone to excitable outbursts but otherwise diligent. She became a patient suitable for discharge.

As a convict prisoner, Margaret's sentence had a defined end date of March 1877. Several years of good behavior and hard work meant that the Home Office was prepared to consider releasing her early. As she approached the last year of her sentence, the Broadmoor staff began to make enquiries as to who might take care of her. Her brother, from whom she had stolen all those years ago, had remained in contact and occasionally visited her and so he was asked if he might help. He was happy to do so, and to offer her accommodation at his lodgings back in London. Once reassured on that point, her order of license for release arrived from the Home Office, and she was presented with the parchment document, signed and sealed. She was discharged on 9th February 1876. Orange paid her fare from Crowthorne and she took the train to Waterloo, reporting her arrival at her brother's house to the Metropolitan Police.

Despite Margaret's good behavior in Broadmoor, her life outside did not change much. She was unable to keep herself away from trouble and remained a petty criminal. At the time of the 1891 census, she could be found resident in another cell, this time in a police station in Paddington.

Margaret Crimmings was the exception to the Broadmoor mothers, in that she was more of a criminal than a lunatic. When it was time for the next baby to arrive, it came from more typical stock. By now, it was 23rd February 1873. A second girl, christened Elizabeth Margaret, this child was born to Margaret Davenport, a thirty-one year old housewife from Warrington, Lancashire. Like Catherine Dawson, Margaret Davenport had also been detained in Kirkdale Prison, and was transferred from there to Broadmoor on 26th September 1872, when she was four months pregnant. She had been detained in Kirkdale a little over two months while she awaited the move.

Margaret had already given birth to four children, including two daughters. These were all now deceased. The two boys had died from natural causes while in infancy; her younger daughter, also Elizabeth, was twenty-two months old, and elder daughter Margaret, six, when in June 1872 their mother had held their heads under water in a pan mug until they drowned. Margaret Davenport had then attempted to drown herself in the tub, then to hang herself, and finally to cut her wrists but had been unsuccessful in all these tasks. So she washed the children, laid them out in her bed and then made dinner for her husband.

She had been found insane when she was due to plead at her trial at the Liverpool Assizes. The supposed cause of her illness was given on her admissions statement to Broadmoor as 'family troubles'. She had married Joseph Davenport in 1862, after they met while working as servants for a landed Cheshire family. Joseph worked long hours as a delivery man, and the family lived a basic existence in the centre of an industrial town. Margaret had apparently been taken ill after the birth of the first Elizabeth, becoming depressed and twice being found wandering the streets at night. The local Police felt that she was the victim of domestic neglect, and that it was her isolation as the homemaker which had led to her depression. She was advised to return to her native Shropshire for a break, and the effect of this was beneficial. A cheerier woman returned to Warrington, and life for the Davenports carried on much as before. There had been no recovery, though, and Margaret was still thinking irrationally. At her first committal hearing after the murders she had stated that 'I was very much provoked before I did it. I was made in hell.'

Now that she was resident in Crowthorne, her mental state continued to be a cause for concern. Like Catherine Dawson, the Broadmoor doctors did not let her nurse her baby. They considered it unsafe for her to do so. Instead, little Elizabeth was taken from her mother at birth, and reared on cow's milk elsewhere in the Asylum. It is unclear who decided to name the girl, and to create the arguably morbid situation where she was named after her dead sisters. It is possible that it was Margaret, for she was a little more reliable than Mrs. Dawson. She saw the baby frequently, though under supervision, and this bonding did not include any unfortunate incidents. Nevertheless, the doctors noted that on more than one occasion, Margaret expressed the hope that her new daughter would die. It would never be safe to let her have the connection enjoyed by Mary Meller or Margaret Crimmings.

In line with previous practice, the Broadmoor authorities busied themselves organising who would take in the child. As Margaret was married, Dr Orange's first correspondence was with her husband, Joseph Davenport. He wrote to Davenport in early April, but the working man refused point blank to have his baby daughter, saying, like Henry Dawson, that he was too poor to be able to take charge of a child and provide care for it. His circumstances were different to those of Mr. Dawson, however, who was already looking after his other children in reduced accommodation. Nevertheless, for the time being, Orange changed his line of enquiry. Instead, his next move also echoed that of the Dawsons' case. He wrote to the Poor Law Guardians for Warrington Union and asked them to take charge of the child instead.

Unlike the Chorley Guardians in the earlier case, the Warrington Guardians did not see their acceptance of the child as the logical outcome. Replying to Broadmoor in May 1873, they stated that they saw no reason why the able-bodied Joseph Davenport could excuse himself from the care of his only living child, and no reason why the burden of her care should fall upon the parish ratepayers. They dared Orange to provide a legal authority upon which he could base his request.

Dr Orange did not give up easily. He saw no benefit to anyone in having the child remain at Broadmoor longer than necessary, and felt that the Guardians of the Union were being unnecessarily difficult. He gathered together what precedent he could find, and wrote again to them suggesting that under statute, the child's legal place of settlement was Warrington; that the father was destitute; and that the mother might destroy her child. The Guardians did not dispute the need for safety, but they did dispute the extent to which Broadmoor could rely on laws created many years before its own invention, and they also disputed whether Joseph Davenport was truly destitute. It was known that he was a working man of working age, employed as a carter, and the Guardians stated confidently that a man in this position would be turned away from their own workhouse, should he fall upon it for relief. By extension, they did not see why there was a need for them to provide poor relief to his child. The Guardians finished off their financial reasoning with an attempt to reclaim the moral high ground, arguing against the harm that could be caused by the removal of such a young child from its parents.

The Home Office was compelled to make a decision in the matter. In July, it instructed Broadmoor's Council of Supervision, and by default, Dr Orange, to send the girl to Joseph Davenport. Orange wrote to him again. This time Davenport sent a long reply in September, once again pleading poverty, and also saying that he had a bad leg which meant that he was currently out of work. No sooner had the situation appeared clear than it was muddied again. Orange forwarded Davenport's response to the Warrington authorities, saying that as ordered, he would still send the child to its father but would be grateful if the Union could stand by if Joseph Davenport refused to take custody of his daughter. The last thing that he wanted was to send an attendant and the baby all the way to Warrington, only to find no room at any inn. He also threatened Joseph Davenport with legal proceedings if he did not agree voluntarily to the arrangement. This threat seems to have finally done the trick. In late October 1873, when she was eight months old, one of the female attendants took Elizabeth on the long journey to Warrington and delivered her to her father.

But this was not to be a happy ending, like the Mellers' tale. Elizabeth Davenport the second was another sickly child, and she would only live for another two years, dying as a toddler at the end of 1875. Joseph Davenport lived on, alone, though he remained in regular contact with his wife down south. He died fourteen years later, in June 1889.

Margaret continued to be a Broadmoor patient while her family's story was played out Warrington. She remained delusional and persecuted. She stated that the other patients threw knives at her, and that she was visited and tormented by them at night, with one particular patient taking the form of a serpent. She evidently lived in fear and tried to hide. Dr David Nicolson, Deputy Superintendent, wrote that 'when spoken to she covers her face with her hand, shuts her eyes and looks downwards and away from the speaker, with an air of intense timidity and shyness'.

By January 1890 Dr Nicolson, then Superintendent, was of the view that Margaret could be discharged to an ordinary asylum. For several years she had been withdrawn and uncommunicative but otherwise well behaved. The official description of her was 'demented' but 'harmless'. It was decided to move her to the Rainhill Asylum in Liverpool, where Catherine Dawson had stayed some three decades before. By now, her husband was dead, and the move north would not bring her closer to any family connections. But perhaps that was irrelevant, as she continued to write to Joseph and to talk to him long after his death. So on 10th February 1890, she was transferred to what became her final home.

At Rainhill, Margaret carried on much as she had done at Broadmoor. She wrote to Joseph and worked a little on the wards until her health failed. For the last seven years of her life, she was effectively immobile. She died on 3rd February 1912, choking on her own vomit as she tried to digest her lunch.

Those four cases in nine years constituted the initial glut of Broadmoor babies. Afterwards, there were fewer cases, and as these drift later towards the twentieth century, a number of the Victorian babies become part of case files which will remain closed for some years to come. There is one more baby to include at present, and this one came after a gap of nearly six years from the birth of Margaret Davenport's child.

This time, the labour was long, despite it being the mother's fourth child. The new baby entered the world at eight o'clock on the morning of 14th January 1879. A third Broadmoor boy, William, he was born to Catherine Jones, a thirty-three year-old farmer's wife from Llanllyfni, Caernarvonshire. Catherine was described by Dr Orange on her admission notes as 'of respectable appearance but with a decided air of melancholy'. She had been brought from Carnarvon Prison the previous September, where she had been in custody since May. She had been aware of her own pregnancy while in prison, and when her transfer was arranged she had informed the authorities that she was pregnant, so they had been prepared for the birth since her arrival.

Catherine's case was yet another of infanticide. She had killed the youngest of her children, her eighteen month-old daughter Sarah. Catherine's was considered by the medical men to be a classic case of 'puerperal mania', or of dangerous postnatal psychosis. She had already attempted to cut her daughter's throat at the family farmhouse in North Wales, when on 9th May 1878 her husband William left her alone with the child in the kitchen for a few minutes. On his return, the child was dead, with blood trickling from its nose and ears. Catherine said that the little girl had fallen from a chair, but her past history meant that this story was challenged. Later the same day she confessed to one of her servants that she had placed her hand over the toddler's mouth until she had suffocated. Her case proceeded to a full trial at the local Assizes, where the jury acquitted her on the grounds of insanity.

Catherine brought an additional complication to Broadmoor as well as her pregnancy. For she could not speak, read or write a word of English. She was a native Welsh speaker, with no other languages. This was a comparatively unusual situation for the Asylum. There were a few patients in Victorian Broadmoor for whom English was not their first language, but many of these spoke French or German instead, and the medical staff, not least Orange, were able to converse in these other tongues. This would not be so with Catherine. When she arrived at Broadmoor, she could not communicate with any of the staff, and so some other method was required. As luck would have it, there was another Welsh female patient who could speak a little of the language, and so she was drafted in to act as Catherine's translator. This was just as well, as Catherine quickly fell ill, showing signs of pleurisy, and was confined to bed.

William Jones was informed of his wife's dangerous condition, and visited her for a brief spell in late October 1878. He too spoke no English and arrived with a handwritten note prepared by friends. This note introduced him to the Broadmoor staff, and asked whether they could recommend him lodgings during his visit. Of course, they obliged.

The fact that no one could understand Catherine was a source of concern to both the Broadmoor doctors and the Home Office. It was not safe to have a patient sick in bed, yet unable to communicate their needs. Orange soon began agitating for his patient's transfer back to a Welsh-speaking asylum, as quickly as her health was up to it. Even when she rallied, after December 1878, Orange still sought to transfer her to an asylum nearer her home before she gave birth.

The Home Office took a different view, possibly as it was so soon after her verdict and sentence had been delivered, and instead asked Orange to find 'some respectable woman, who can speak the Welsh language' to act as a dedicated attendant to Catherine. Orange retorted that employing a dedicated member of staff to act as translator was not seen as practical. So the other Welsh patient, who came from Glamorganshire, continued to act as Catherine's official interpreter during her time at Broadmoor.

Perhaps because of the inability to communicate with her, the staff at Broadmoor did not feel able to let her nurse her child, and the baby boy was removed from her immediately after birth. Without the possibility of a thorough interview, and given her previous medical history, it was felt too risky to leave little William in the sole care of his mother. One of the female attendants, Harriet Hunt, took charge of him instead. The suggestion, though, is that Catherine was recovering from her mental illness, even if her physical health continued to be poor. She was allowed to see her baby in the infirmary, and bond with him while the usual arrangements were made for his removal. This case was a simple one, as William Jones was very eager to take care of his infant namesake. He visited both mother and child regularly before he took the three-month old baby home on 16th April 1879, with Harriet Hunt, the nursemaid, accompanying him on the journey.

At roughly the same time, the Home Office finally acquiesced regarding Catherine's transfer. They delivered the warrant that Dr Orange had requested to remove Catherine to the Joint Counties Lunatic Asylum at Denbigh in North Wales. Yet Orange's satisfaction was tempered by the fact that Catherine's health took another turn for the worse. She was bedridden again, and her transfer was postponed. Over the spring, she remained in Broadmoor's infirmary while her husband and child were at home.

Fortunately, this experience was to be short lived. As before, she rallied, and by July she was sufficiently well enough for her transfer to be effected. Orange wrote to Denbigh, and a female attendant from that Asylum arrived by train on 29th July 1879 to collect Catherine and escort her back to Wales. She had stayed in Broadmoor for a very short time, a little over ten months, but for the time being she remained a pleasure woman.

The care that Catherine had received in Broadmoor had been considerable, and this was acknowledged by her family. The last paper on her Broadmoor file is a letter written on behalf of William Jones in January 1880. In it, he stated that although his wife seemed rational and sane in Denbigh, her general health was worse, and he ascribed this to the inferior diet she was given compared to her Broadmoor rations. He asked for Dr Orange's help in gaining his wife's discharge back home.

The Home Office relented in her case within a year. She was conditionally discharged from the Joint Counties Asylum and moved back to the farmhouse that she now shared with William senior, William junior and the other children at Llwydcoed Fawr in Llanllyfni. Her husband carried on with the farm, and she carried on as a mother, that day in May 1878 now forgiven, if not forgotten.

These women's stories are only five of some five hundred from the Victorian period, but they serve as an illustration of the type of case to be found in Broadmoor's female wing. Apart from their confinements, these mothers blended in amongst the other women on the wards. Their crimes were unremarkable, even if we find them shocking.

There is no evidence that the medical staff at Broadmoor ever sought to follow up the fate of the children who had left their care. Any subsequent discovery was down purely to communication from outside. As it turned out, the Broadmoor babies had suffered differing fortunes. The poor law welfare system had intervened for three of them, which perhaps says something about the social class of woman likely to be found in the Asylum. Only two ended up being cared for by their own families.

For the babies that lived, by the time they arrived in adulthood they would have had no recollection of the place where they had spent their first few weeks of life. They would not recall the walls, the wards or company of lunatics. It is unlikely that they considered themselves to have been born in Crowthorne. The fact that the hospital had no further business with them meant that they were also free to make their own lives away from any taint or stigma. Their stories would remain separate from that of Broadmoor until now.

Sent to Broadmoor

Every patient had been arrested for a crime, and then dealt with by the courts. Most of these had been found 'not guilty by reason of insanity' (at least until 1883, when the standard form became 'guilty, but insane', in a vain hope to deter lunatics from their actions by denying them innocence), just as James Hadfield had been so found in 1800. These were the 'pleasure' men and women, destined to remain in Crowthorne until what was now Her Majesty's Pleasure was known. Although the balance varied, roughly two-thirds of the patient population at any time were 'pleasure'. These patients had been declared insane usually at their trial or before it. Some patients did not even get as far as making a plea, while others were found insane on arraignment, when they came to stand in the dock, but before any evidence was heard. If a case went to full hearing, the jury would have delivered a verdict of insanity based on the evidence put forward, usually by the defense. The yin to this yang were the 'time' patients. These criminals were all guilty, but not initially insane. After their conviction, they had been given a custodial sentence by the courts, and became prisoners. Sentence length varied: most convicts at Broadmoor were serving somewhere between five and ten years, though their number included murderers who were serving a life sentence, commuted from their appointment with the gallows. The usual passage into Broadmoor for the convict patient was that during their sentence they had become insane, and therefore in need of treatment in an asylum. Those with lesser sentences tended to be farmed out to the county asylum network, with Broadmoor reserved only for the more truculent types. The second way in, somewhat rarer, was that they faced the death penalty, and the Home Secretary had ordered a special inquiry into their sanity. A number of murderers were respited to Broadmoor's care in this way. Usually they retained the guilty verdict, such as Mary Ann Parr; exceptionally they might become an innocent 'pleasure' patient instead, such as Christiana Edmunds. This escape route from the clutch of death (or even incarceration) might beg the question of whether any fake lunatics were to be found within the walls. Evidence exists that suggests the possibility arose, though also that an attempt to feign mental illness was often without success. Broadmoor's staff were wise to the possibility of malingerers, and there was a revolving door that returned as many convicts to the prison system as it received; quite apart from which, a sane convict soon discovered that sharing space with the lunatics was not necessarily preferable to the greater rationalities of jail. The more intriguing question to consider is whether any sane murderers cheated the noose. This is an investigation that also reveals a time when mental illness was understood rather differently to how it is today. For the lunatics were, by definition, insane. Though they were no longer diagnosed as being affected by the moon, they were affected by things that did not so affect the other, non-lunatic people. There was an element of mystery about their disease, something intangible about how it made effect upon their bodies. The word 'lunatic' has itself become a somewhat guilty word of late, an incorrect way of describing a sufferer from mental illness. This seems a shame: the word is ripe for reclaiming by those afflicted by the moon. It is a word of great power, and potentially empowerment. It aptly conveys the loss of control and influence over one's actions to forces both outside our control, and not fully understood. The Victorian definitions of insanity were different to our own, though they recognised the same phenomena. I have already written about the idea of 'moral' and 'physical' causes, something which only began to die out as the nineteenth century drew to a close. As far as the doctors were concerned, these causes then manifested themselves in defined diseases, each of which might be inferred by observing the patient's habits, as well as through interview. These diseases are still recognisable today: mania, melancholia, and dementia. Monomania was an obsession with a single subject; amentia, absence of mind, would be described as learning disabilities, now recognised as something completely separate from mental illness. To these cognitive deficiencies, the Victorians added the concept of moral insanity. This was a disease free of delusions, but where the mind was unable to think and behave properly as it should. Although it did not fit the modern term of psychopath, itself a rather overworked word, it is perhaps the nearest to it that the Victorians acknowledged. Of course, for all these diseases, it was not sufficient to merely be a sufferer for a plea of insanity to succeed: the defendant's legal team also had to show that the disease had led to the action, and that consequently any means was absent.

So it was these patients who were given the 'pleasure' sentence. They either stood in court, or did not even make it that far, while legal argument was had as to whether or not they were culpable for their actions. The basic rules covering the insanity defense were laid down by the McNaughten Rules in 1843. Daniel McNaughten had killed the private secretary of Sir Robert Peel, almost certainly in mistake for the Prime Minister, and then, far worse, upset the popular press by being found not guilty for the crime by reason of insanity. It took the entire House of Lords to deliver guidance that effectively confirmed the correctness of McNaughten's verdict, and guaranteed that he was spared the noose. McNaughten ended up first in Bethlem and then in Broadmoor while his Rules lived on. The most-quoted premise from McNaughten was that the defendant was unable to reason right from wrong, and so did not understand the nature or the quality of his or her actions. It was a fine judicial statement, at once precise and yet still leaving plenty of room for legal argument, so the lawyers undertook their increased scope for discourse with enthusiasm. Various approaches became popular: showing that your client suffered from particular delusions was one, often linked to some sort of traumatic event, past or present. A destitute man may believe his family better off in heaven, or a new mother that her child was permanently blighted by sin. Similarly, the insane actor may be driven to his crime by an irresistible impulse, at the mercy of forces beyond his control. Drink, if taken to addiction, could effectively cauterize choice. The casual observer might well conclude that the law was drawn more generously than it is today. An alcoholic is unlikely to be found not guilty, and the perpetrator of crimes that we find it difficult to understand is no longer likely to be given any benefit of mental doubt. Yet many of the celebrated insanity cases concerned murder, and the law of the Victorian court had a heavier weight to balance on its scales of justices: that of the condemned's feet upon the gallows trapdoor. Perhaps the law is only human, after all. Having been defined by the courts, or a prison doctor, as suffering from one of these diseases, a patient was transferred to Broadmoor to begin their 'moral treatment'. As mentioned before, the routine of patient life was an integral part of their care, and it is worthy of further exploration. Routine would be a feature of every life within the institution, though the nature of the routine was itself subdivided. This division began when a patient was assigned to one of the Blocks, as each block was quite separate, and segregated. On the female side, the initial Block housed all the patients. There was a divide between three wards: one ward for the more aggressive or noisy patients, one ward for those who were low risk, and one ward for those in-between. When the further block was opened in 1867, the more aggressive females were siphoned off into that. This picture was mirrored on the male side on a grander scale. By 1868, the full complement of six blocks was complete. Blocks 1 and 6 were known as the 'back' or 'refractory' blocks, for dangerous and violent patients. The men here had their own separate airing courts, bricked in and hidden from the rest of the site, and the attendants who tended them wore uniforms with padding and with hidden buttons. The name 'back Blocks' came from their position, which was on the north side of the site and away from the beautiful views across the southern Terrace. The back Blocks contrasted with those nearest to the terrace and the wider grounds, which were Blocks 5 and 2. Patients in these blocks were considered the lowest risk, and enjoyed greater access around the site. Block 2 in particular became known as the privilege block, where patients had the most freedom to plan their day. Their insanity did not affect their daily lives, and they could be trusted to spend their time fruitfully at work, in their rooms, in the communal rooms in their block, or on the terrace. Block 2 was where VIPs and the press were brought if a bit of Victorian PR work was required. Oxford, Dadd and Minor were all sometime residents of Block 2. Block 3 housed the infirmary, and Block 4 included the admissions ward, but both these blocks also housed those in-between patients who did not fit into the categories of being either dangerous or trustworthy. These were the biggest blocks, housing one hundred patients each, and also had the most communal dormitories on the site. Dormitories were gradually reduced in number during the Victorian period, with the result that the majority of patients had a single room to themselves. Such rooms measured twelve feet long by eight feet wide, and were equipped with a single bed, or a mattress only in the back blocks, and a desk. The linen was changed twice a week. Patients were also allowed personal possessions if it was safe to have them, which would vary from patient to patient and block to block. A set of cufflinks proudly worn in Block 2 would become a potential weapon in Block 1. Once assigned a block, a patient could settle into his or her routine. That would mean a day which started at 6am (or 7am in the winter), when the day shift attendants came on duty, and ended at 7 o'clock at night when the night shift came on. In between those fixed hours, the day was punctuated by segments of time filled by meals, work and leisure. The bulk of the day would be spent at work, if a patient was able to do so. For those capable of only basic labour, work consisted of ward cleaning, the endless washing, scrubbing and polishing required to keep the Asylum and its contents clean. For the more able, women were employed as seamstresses or in the laundry, and men as tailors, shoemakers, upholsterers, tinsmiths or carpenters, or on the Asylum farm, garden or wider estate, tending crops in the fields. Victorian Broadmoor was a largely self-sufficient community, and much of the patients' work benefited directly their quality of life. Such leisure time as there was might be spent reading or playing games in the day rooms in each block, walking in the airing court attached to the block or, for the more trusted patients, playing outdoor sports such as croquet or bowls or even walking (accompanied, of course) around the local area. Evening entertainments were regular, though not frequent, and cricket was played in the summer months. Special interests were encouraged, such as Dadd's painting or Minor's research work. Despite these spiritual comforts, physical comfort could be hard to come by. A patient's life could be cold and dark. At first there was no heating in any of the bedrooms, with only open fires and hot air grates in the day rooms to provide any warmth. Central heating was slowly introduced to the blocks from 1884, first through solid fuel and then by gas, though it was still a while before the individual rooms all felt the benefit. Similarly, oil and gas lamps were used for lighting the communal rooms and corridors until the end of the nineteenth century, but there was no artificial lighting in the patients' bedrooms. In the winter months, patients spent half the day in darkness. Patients changed their clothes at least twice a week, were washed daily, and bathed once a week in the block's bathroom, under the careful eye of an attendant. The male patients were also shaved by an attendant, if they wished to be. Such was the risk attached to this operation that while one attendant worked the razor, another attendant was always present to keep an eye on proceedings. Patients were fed four times a day. Everyone was returned to their block to be fed, as each block had a dining room for its own use. Before each meal, every item of cutlery was counted out by one of the attendants, and then counted back again at the end of it. Although diets varied, it is possible to describe a basic pattern of food. For breakfast, patients generally had tea, and bread and butter. Lunch was bread and cheese. In the early evening, a typical meal would be mutton, beef or pork with potatoes (or vegetables if in season), followed by a steamed pudding. Three-quarters of a pint of weak beer might be given with the evening meal, though further rations of beer were usually given to workers during the day, and brandy or other fortified drinks might be offered to those suffering from physical debility. The final meal was supper, which saw the offer of a further helping of bread and butter with tea. Charged with implementing this routine was a staff of around one hundred Asylum employees. Two men were there at the start: Medical Superintendent John Meyer, and his Deputy, William Orange. They recruited a third doctor as well as the much greater number of male and female attendants, who were the bulk of their employees, and provided the nursing staff in Victorian Broadmoor.

The attendants often had little or no previous medical background, and physical presence was considered as important an attribute as any other. Many of the male staff had either served in the forces or come from the prison service to join Broadmoor's establishment. The early years, in particular, saw a mixed success with this recruitment strategy, as in the 1860s the annual rate of turnover approached 50%. It was expected that female attendants would resign upon marriage, but discipline was also a significant problem. The Asylum archive includes staff 'defaulters' books' that list dishonesty, incompetence and drunkenness amongst the attendants' sins. It would be wrong though, to conclude that this was an inhumane regime, where brutality and immorality were commonplace. On the contrary, there were a number of rules in place which provided attendants with both a moral compass and with procedures for physical restraint. The latter was seen as a last resort and all incidents tended to be noted in one record or another. The large turnover of staff gradually decreased as well in the period after 1870, when Orange succeeded Meyer. The Asylum appears to have been a happier place under Orange, and amongst other things he made small improvements to the terms and conditions of the attendants' employment. Perhaps he also leant a different touch to recruitment. The personality of Broadmoor's chief doctors was bound to leave an impression on the institution that they ran. There is a little more about Meyer and Orange in the Escape from Broadmoor chapter to give you an outline of each doctor's character. It is possible to cast Meyer in a slightly more villainous role: a man who seems to have fought with most of his senior staff at one time or another; a man who had the most violent male patients segregated in caged areas of their blocks; a man who perhaps was not the most enlightened brain doctor of the Victorian age. Nevertheless, Meyer had the unenviable task of trying to find a blueprint for a new type of institution, and also dealing with the inevitable flaws in the design and fabric of the building he inherited. He was nearly fifty when he took charge of Broadmoor, having previously run the Convict Lunatic Asylum in Tasmania, served in the battle hospitals of the Crimea, and then led the Surrey County Asylum for a period before he was charged with mastering Broadmoor. He also suffered from ill health. He was attacked by a patient called John Hughes in the Asylum Chapel in March 1866, struck a severe blow on the temple by a large stone, and never fully recovered. Hughes, a despoiler of holy images in a north London church, stated that Meyer had accused him of 'murdering the Queen of Heaven', and that he was obliged to avenge that insult. He was put in solitary confinement for his trouble.

Attacks would form a part of each of the first three Medical Superintendents' careers, and were an occupational hazard. Orange was attacked by an insane cleric called Henry Dodwell in 1882, who argued that attacking the Superintendent was the only way to draw attention to his wrongful detention, much like he had argued a few years before that shooting at the Master of the Rolls was the only way to draw attention to the injustices of a legal action he was pursuing. Orange's successor, David Nicolson, was similarly attacked by Henry Forrester in 1884 while employed as Deputy Super. Nicolson was well enough to return to work and take promotion in due course, though he was also the only Superintendent to suffer two attacks, after James Lyons went on to throw a stone at his head in 1889. Despite these twin assaults, Nicolson might still consider his to be a more fortunate outcome than that of the Deputy he had in turn succeeded: William Douglas lasted all of four months at Broadmoor in 1871 before patient Henry Leest injured him so badly that he never returned to work.

When Meyer died suddenly in Exeter in May 1870, while returning from a visit to his dying brother-in-law, it was his thirty-seven year-old assistant who succeeded him, and spent the next sixteen years in charge. William Orange is a fascinating character, and it is no surprise that the fine portrait of Richard

Dadd's 'Broadmoor officer', which hung in the Superintendent's office at the Hospital until the turn of the twenty-first century, has been historically attributed as Orange. In terms of this brief introduction to Broadmoor, Orange's importance is the cultural mark that he imprinted onto the Asylum, echoes of which are still apparent today in the twin pillars of rehabilitation and public protection that Broadmoor represents. In that any long-running institution bears a received memory and received values from those who have trod its corridors along the years, it is to Orange, and to Nicolson, that I feel the modern hospital still owes a debt. Orange's care for his staff has been mentioned; from his patients, comes testimony of genuine warmth that still litters the archive. Two personal items might serve to illustrate that: that he received spontaneous letters of goodwill after Dodwell's attack on him; and that Henry Leest, the beater of poor Dr Douglas, felt able to write asking Orange for a little money many years after his discharge. Orange usually obliged his ex-charges with a small sum to tide them over, and there is no reason to suppose that Leest was an exception. Orange was severely incapacitated after Dodwell's attack, with the result that Nicolson gradually assumed more control after summer 1882. When Orange finally retired in 1886, as for the end of Meyer's reign in 1870, it was his Deputy who took over. The third Medical Superintendent had been on the staff since 1876, and remained a personal friend of Orange as the latter enjoyed a long retirement. Indeed, Orange even returned to the Asylum as a member of its scrutiny body, the Council of Supervision. Nicolson provided continuity, as well as a more strategic approach to management than Orange, only ever criticizing his friend and former boss for his micro-management, feeling that at times Orange's attention to detail was not appropriate. However, although my impression of the Orange and Nicolson years is one of great success in their enterprise, when the time came for Nicolson to retire in 1895, his Deputy was not selected to succeed him. The doctor in question, John Isaac, was as old as Nicolson and not quite the high-flyer that his bosses had been, having pre-dated Nicolson at Broadmoor. Instead, the post was given to the suitably-named Richard Brayn, the last of the Victorian Superintendents. Brayn came from the prison service, rather than a medical background, and despite his popularity with the politicians outside the walls (he gained the knighthood which would never come to Nicolson), his period in charge was one of greater tension inside them. Brayn was a great believer in running a tight, disciplinary ship, which occasionally put him in conflict with other professionals around him. The result was that the pillar of rehabilitation was perhaps slightly shorter than the pillar of public protection during Brayn's time in charge: the positives in the lopsided emphasis being a lack of successful escapes, coupled with Brayn's success in becoming the first Superintendent not to suffer personal injury. He was a competent leader who brought Victorian Broadmoor into the twentieth century, and was well-respected by his peers, even if perhaps the same affection for Orange and Nicolson did not extend to him.

Back to Bedlam

Between 1866 and 1869 the Governors of Bethlem built a convalescent home at Witley, designed by Smirke, for patients on the way to recovery, but the main part of the institution remained in Southwark until after the 1914–18 war, when the Governors decided to build new premises in rural surroundings. The removal to Monks Orchard at Addington in Surrey was sanctioned by Act of Parliament in 1926. The freehold of the old site was purchased by Viscount Rothermere in 1930 and vested in the London County Council for the formation of a public open space, to be known as the Geraldine Mary Harmsworth Park in memory of his mother. The side wings and some other parts of the building were demolished. The central portion of the front, with the dome looking disproportionately high above it, and the rear galleries were leased to the Commissioners of Works to house the Imperial War Museum.

The Beginning of Insanity.

"The inmates are ghosts whose dreams have been murdered" Jill Johnston, U.S. journalist after she observed "patients" in New York's mental ward at Bellevue Hospital. Throughout medieval times in Western civilization, people who displayed any sign of mental illness were treated with fear, revulsion and often times, violence. The "treatment" of such people frequently consisted of simply locking them up in a dungeon and ignoring them. They were considered possessed by demons or the devil. Many were murdered or burned at the stake, victims of a misdirected religious fervor that claimed thousands of victims, especially during the Inquisition in 13th century Europe. Organized crusades against so-called "heretics" were formed by the Roman Papacy to persecute those of lesser faith. Soon, the Inquisition was used to severely punish political enemies, criminals and the mentally ill. In England, pressures on the ruling classes, forced them to deal with those who appeared "different" than others. A place was needed to treat and house the mentally retarded and others with mental afflictions that could not be explained. As a result, the Priory of St. Mary of Bethlehem Hospital in London opened its doors in 1247 Encarta 2000. The insanity defense has its roots firmly embedded in centuries of legal tradition. As early as the 13th Century, the English Lord Bracton established the principle of mental deficiency in human behavior. He said that some people simply do not know what they are doing and act in a manner "as to be not far removed from the brute" (Menninger, 1968, p. 112). From that concept, "insanity" came to mean that a person lacks the awareness of what he or she is doing and therefore cannot form an intent to do wrong. Since there was no malice in the intent of his or her actions, then there could be no technical guilt. The standard for insanity in the courts was determined to be such that a "man must be totally deprived of his understanding and memory so as not to know what he is doing, no more than an infant, brute or a wild beast" (Melton, 1997, p. 190). This "wild beast" standard was the insanity requirement of England's courts for over a hundred years and any defendant who attempted to use the defense had to prove he or she lacked the minimum understanding of a wild animal or infant. It wasn't until 1843, when a man named Daniel M'Naghten committed a murder that would alter forever the history of jurisprudence in the Western world. In 1843, Daniel M'Naghten, a Scottish woodcutter, shot and killed Edward Drummond, secretary to England's Prime Minister Sir Robert Peel in London. He acted under the belief that he was actually shooting the Prime Minister because M'Naghten believed there was a plot against him. When M'Naghten reached trial, his attorneys pleaded that he should be acquitted because he was obviously insane and did not understand what he was doing. M'Naghten was later acquitted of the crime. Later that same year, the House of Lords issued the following ruling: "To establish a defense on the ground of insanity, it must clearly be proved that, at the time of the committing of the act, the party accused was laboring under such a defect of reason, from disease of the mind, as not to know the nature and quality of the act he was doing; or if he did know it, that he did not know he was doing was wrong" This edict became know as the M'Naghten Rule and for over a century, this was the standard for the insanity defense. As for Daniel M'Naghten, after his acquittal, he was sent to Bedlam and other institutions where he languished in the shadow world of the insane for several decades until his death in 1863. The late 19th century was a time when scientific ideas were rampant. This explosion of science was partly brought on by the publication in 1859 of one of the most influential books ever written, The Origin of Species by Charles Darwin. The Darwinian concepts of survival of the fittest, natural selection and hereditary traits revolutionized biological science and were applied to many other disciplines. Ideas were evolving rapidly during this era in every medical, scientific and psychological field. Things were changing quickly in the legal profession as well. Dedicated lawyers and judges searched for workable solutions to the controversies that plagued the nation's courts. The confusing ideas about mental diseases and the complexities of the human mind did not lend themselves well to the rigid dimensions of codified law. A glaring example of that confusion was the murder trial of Charles Guiteau in 1881, assassin of President James Garfield.

President James Garfield

Guiteau was an erratic individual who suffered from some type of mental disorder, most probably paranoia. He had delusions that he should be appointed as Ambassador to France because he wrote a speech for President Garfield which he imagined helped get Garfield elected in 1880. The speech, in fact, was never used but Guiteau became despondent and bitter over this "betrayal" and plotted to get revenge against Garfield. After stalking the President for almost a month, he managed to shoot Garfield in the back at a Washington D.C. train station. The President was not killed outright however. He suffered with great pain for almost 3 months before he finally died on the night of September 19, 1881.

Charles Guiteau

Guiteau's trial was a sensation, not only because of the nature of his crime but his unusual outbursts and behavior in court attracted widespread attention. Guiteau testified in his own behalf and stated: "I want to say right here in reference to protection, that the Deity himself will protect me, that He has used all these soldiers, and these experts, and this honorable court, and these counsel, to serve Him and protect me" (Knappman, 1994, p. 190). He went on to tell the court that he shot the President because God told him that Garfield was destroying the Republican Party and he must die to save the Democratic Party. But the prosecutor, U.S Attorney Wayne Davidge told the jurors: "It is very hard to conceive of the individual with any degree of intelligence at all, incapable of comprehending that the head of a great constitutional republic is not to be shot down like a dog" (Knappman, 1994, p. 190). Charles Guiteau was found guilty of murder on January 13, 1882. Upon hearing the verdict Guiteau jumped to his feet and screamed "You are all low, consummate jackasses!" (Knappman, 1994 p.190). After reciting an incoherent poem in a child's voice on the gallows, Guiteau was hanged on June 30, 1882. Although there was a strong public sentiment to punish Giuteau, despite his severe mental problems, his case underlined the need for change on how the courts dealt with the issue of insanity.

Christmas at Bedlam

☐Bethlem Hospital in the eighteenth century which is not as well known as perhaps it should be: that admission was commonly for a period of no longer than twelve months. What was true of the Georgian and Regency Hospital in Moorfields also held good for the Victorian and Edwardian Hospital at Southwark. There were, however, always a few exceptions that proved the rule people who stayed longer than twelve months especially after the establishment of the Hospital's incurable ward. Emma Lane was admitted in May 1893 after having spent twenty years of savings in a matter of weeks on unneccesary food, baby clothes and theatre bookings. Her husband kept in close contact with the Hospital throughout her extended stay, at one stage writing 'I am anxious to see her resume her old place, but fear she is not yet well enough'. Emma was granted temporary leave to spend time with her family a number of times, including at Christmas 1893 and 1894, but matters did not run smoothly, and on each occasion she was returned to the Hospital. Christmas 1893 seems to have been particularly stressful, the family's report being that Emma had been 'giving trouble', Emma's version of events being that she had 'just bought a few things'. Emma was discharged uncured in January 1895; the story of her hospital stay may be read in Presumed Curable: An illustrated casebook of Victorian psychiatric patients in Bethlem Hospital by Colin Gale and Robert Howard (Wrightson Biomedical, 2003). Of course, not everyone could be sent home for Christmas, and the Hospital's second strategy to maintain seasonal morale seems to have been to bring Christmas to the majority of patients and staff that remained in residence throughout. The photograph below offers remarkable evidence of one attempt to do so. It shows a statue that stood in one of the Hospital's galleries (shared ward space) dressed as St Nicholas for the Christmas season of 1907. To our contemporary gaze, the visual effect is unusual, even a little unsettling. Yet the intention must have been to lift the spirits, and we may hope that the display succeeded in doing so at the time. In any event, all the staff of the Archives & Museum wish the readers of this blog a very happy Christmas and safe and prosperous New Year.

Under the Dome: Open House London at the Imperial War Museum

The large green dome above the main façade of the Imperial War Museum has been one of the most distinctive features of the building since it was added to the Bethlem Royal Hospital during improvements completed in 1846. As part of Open House London last weekend, we ventured inside to see how much of nineteenth century Bethlem remained. Now the dome is no longer the Imperial War Museum's reading room (it has been transferred to the fully accessible new Explore History Centre), opportunities to visit are rare. However, 130 lucky visitors (and us!) made it up the three flights of stairs into the dome last Saturday, to hear about the history of the building and the Imperial War Museum from their Archive team, and browse some nineteenth century casebooks: records of patients who may well have attended services when the dome formed the Hospital's chapel. This part of the musium is not usually accessible to the public is the Victorian Bethlem Hospital at the Imperial War Museum. Opened in 1815, when Bethlem was moved from its crumbling former premises at Moorfields, the Hospital was located on this site until 1930, when it moved to its present location in Beckenham. Although the conversion of the building to the Imperial War Museum, established in 1920 and opened on this site in 1936, as well as extensive bomb damage in the Second World War (a total of 41 incidents) means that much of the building's original fabric has been altered, the facade is still distinctly recognisable, while the pathways and walls in Geraldine Mary Harmsworth Park, formerly the Hospital grounds and airing courts, still follow the plans of nineteenth century Bethlem. Most of the former Hospital rooms now form the "behind the scenes" areas of the Imperial War Museum, with the public galleries located in what was originally a central garden. Some of the most distinctive Hospital locations, however, will be open for visitors on Saturday 18 September only, as part of Open House London weekend: the Dome and the Boardroom. Smirke's Dome, added to the Hospital during improvements carried out between 1838 and 1846, was one of the most distinctive features of the nineteenth century building: the patient-edited Hospital magazine, begun in 1889, was titled Under the Dome. The Dome contained Bethlem's chapel, recently, after restoration following an arson attack in 1968, the Dome housed the Imperial War Museum's Reading Room (from May 2010, this was moved to the new Explore History Centre). The guided tour, led by archivists from the Imperial War Museum, will also take in the Boardroom - the only room in the building still used for its original purposes, having formerly served as Boardroom for Bethlem's Governors. The room currently contains a collection of artworks by William Orpen. A visit made to Bethlem by the French socialist and proto-feminist thinker Flora Tristan in 1840. She signed Bethlem's visitors' book (now held in the Archives) and later wrote of her experiences in *Promenades dans Londres* (published in English translation under the title *Flora Tristan's London Journal 1840*). In it she makes the ostensibly unflattering observation that 'it is generally accepted that England is the country with the greatest number of insane'. But an explanation is offered for this: England is, according to Tristan, 'the country where free inquiry gives rise to the greatest number of religious and philosophical sects the more a people is inclined, by its religion and its philosophy, to resignation, the fewer madmen there are in its midst; whereas those peoples who by reason govern their religious beliefs and their conduct in life are those among whom one finds the greatest number of insane'.

Evidence of the room's former use (carefully restored following an arson attack in the 1960s) still remains, including the Ten Commandments displayed on the wall, above where the altar would have stood, and the gallery, which used to house the choir. The Hospital Chaplain was an important part of daily life at Bethlem; as well as providing religious and spiritual counsel for patients (and staff), he was also heavily involved in the programme of entertainments. Rev. Edward Geoffrey O'Donoghue (Chaplain from 1892 – 1930) organised fortnightly "Working Parties," in which female patients were "encouraged to forget their own maladies in working for others." He also wrote a history of the hospital, and gave regular lectures to staff and patients on the topic: over 700 lantern slides he used to illustrate his talks remain in the Archives.

Secure Accomadation For the Crimally Insane.

A Select Committee of the House of Commons had recommended the erection of a national asylum for criminal lunatics in 1807. In 1810, the Governors of Bethlem, who were at that time planning the building of the third Bethlem Hospital at St George's Fields, were approached by the Secretary of State for the Home Department with a proposal to provide secure accommodation for criminal lunatics in a new wing of the building. An agreement was hammered out over the next four years, in which the Home Department would be responsible for the government of State Criminal Lunatic Asylum, and would pay for the erection of the new wing of the hospital in which it would be housed, and for the maintenance and medical care of patients; and the hospital would provide the requisite facilities, and the majority of its medical personnel and care (all paid for by the state). In effect day-to-day control of the Asylum rested with the Bethlem physician acting in accordance with statute. The two criminal blocks were opened in 1816, and criminal patients were kept entirely isolated from patients in the main hospital. Admission to and discharge from the State Criminal Lunatic Asylum was by warrant of the Secretary of State, and quarterly reports on criminal patients were made to the Secretary of State by the Bethlem physician. Returns continued to be made until the 1880s, despite the transfer of the majority of criminal patients to Broadmoor by 1864, and the subsequent demolition of the wing of the hospital in which they were housed, because a small number of criminal patients were kept at Bethlem Royal Hospital.

Broadmoor

"There are many sad tales of lives destroyed by mental illness, of families broken up and never mended, of fear and paranoia." Among many famous patients whose records become available from November 18, is the artist Richard Dadd, who painted many of his best-loved pictures during his time in Broadmoor. A celebrated watercolourist, Dadd suffered a mental breakdown and stabbed his father to death in 1843. He was pronounced insane and committed to Bethlem asylum then transferred to Broadmoor as one of its first patients in 1864. Encouraged by staff, he continued to paint, despite violent intervals. Broadmoor account books detail his orders for peppermints and sable brushes. The asylum was isolated and self-sufficient, with 170 acres of farmland and workshops for shoemakers, upholsterers, tinsmiths, carpenters etc, where the male patients would work. The women were expected to work in the laundry, as well as sewing and cleaning. While the sexes were strictly segregated, their 'therapy' was a routine of work, exercise and rest, with newspapers, games, billiard tables, pianos, and a small library all available to them, as well as a chapel for daily services. The patients would play bowls, cricket and croquet. Broadmoor was managed by a 'Medical Superintendent' who along with two doctors were the only medically trained staff at the asylum, whilst about 100 untrained attendants acted as nurses. They all lived on site and would entertain the patients by putting on variety shows in the main hall. The archive deposited at the record office covers the period from 1863 to 2004, however, due to the 100 year closure rule, only records older than this have been made available to the public. The commissioners later acquired an advisory function, but their primary duty remained to visit all institutions regularly, reporting annually to the Lord Chancellor and to Parliament. In 1853 counties and boroughs were obliged to provide an asylum. The same year the Bethlem Hospital was brought under their supervision. Criminal Lunatics were sent to Broadmoor. Christiana Edmunds is a typical example.

Christiana Edmunds:

The most celebrated Victorian female patient at Broadmoor has been remembered for the cause of her admission rather than any wider social impact. This is perhaps a reflection on how scandalous women fulfilled the voyeuristic delight of Victorian society. For Christiana was a woman who satisfied certain stereotypes, and her story included sex and murder. The tabloids christened Christiana 'The Chocolate Cream Poisoner'.

Born in Margate, Kent, the daughter of a local architect, and sent to private school, Christiana grew up in a household already touched by insanity. For the Victorians, the mental illness found in Christiana's close family would prove to be a strong factor in her own diagnosis. Hereditary insanity was marked: her father had apparently gone mad before his early death, and two of her siblings died in adulthood, a brother in Earlsfield Asylum in London, and a sister allegedly by her own hand. Nevertheless, she came from a very comfortable, middle class background, and was described at her first trial as 'a lady of fortune, tall, fair, handsome and extremely prepossessing in demeanor'. From the age of around fourteen, she lived alone with her sister and their mother, an aging landlady.

Little is known about her early adult life, except that as a party to an independent income, she did not need to work. The family moved to Brighton in the mid 1860s. Her recorded history properly begins when in the middle of 1869 she first met, and then fell in love with a Dr Charles Beard who lived nearby. She sent him love letters, and, to begin with he reciprocated her friendship. In such times, any form of intimacy was significant, and it appears that they carried on some level of romantic relationship for the next few months. The nature of this level has to remain a matter of conjecture, and the extent of the relationship may have been greater in Christiana's mind than in reality. Dr Beard always maintained that there had been no affair in a physical sense, but even if it was purely an emotional affair, some sort of connection had been made.

There was a small problem, however: Dr Beard was already married. He now found himself a respected member of the local community who was being disloyal to his wife. Whatever he was up to, it was unwise. During the summer of 1870, the burden of deceit became too much, and Dr Beard asked Edmunds to stop writing to him: 'This correspondence must cease, it is no good for either of us'. Edmunds did not stop. By now, she was used to calling on the Beards from time to time, and she used this familiarity to take additional action. One day in September 1870, Edmunds visited Mrs. Emily Beard, the good doctor's wife, with a gift of chocolate creams for her. Mrs. Beard ate some of the chocolate, and was promptly, and violently sick afterwards. Dr Beard accused Edmunds of poisoning his wife, although Edmunds refuted the allegation. Instead, Christiana complained that she was as much a victim as Mrs. Beard, for the same chocolates had made her sick too. Beard withdrew his accusation, but Edmunds was banished from the Beard household, after a last, climactic meeting in January 1871. Dr Beard also wished to banish Edmunds from his life, but in this respect he was not successful. The letters continued to arrive at his offices, sometimes forwarded to him from home, two or three times every week. He ignored them.

This might have just become another case of a spurned lover, except that over the next few months there were many further cases of people falling ill in Brighton after eating sweets and chocolates. None of these cases was newsworthy on its own, despite their personal drama. All of them featured a violent sickness, which passed quickly and without lingering harm. Consequently, stories of them spread by word of mouth rather than through the local media. Then on 12th June 1871, a man called Charles Miller, on holiday in Brighton with his brother's family, bought some chocolate creams from a sweet shop called J.G.Maynard's, ate a few, and gave one to his four year-old nephew, Sidney Barker. Miller became ill but recovered. Barker died.

This was altogether a more serious episode. It was necessary to hold an inquiry into the tragic event. Amongst those who came forward to give evidence at the inquest was Christiana, who claimed that she and her friends had also become ill after eating sweets from Maynard's store. She blamed Mr. Maynard for some personal discomfort caused the previous year, when the wife of a good friend had suffered a similar event. There was evidence to back this up, because tests before the inquest discovered strychnine in the chocolates sold by Maynard's. What was not resolved at this inquiry was how the strychnine had come to be within the chocolates. As a consequence, a verdict of accidental death was recorded on the boy, and the shop owner John Maynard exonerated of any intentional poisoning. He destroyed all his stock.

If, at the time, Barker's death was considered to be an unfortunate accident, there followed a series of occurrences to arouse suspicions of foul play. Shortly after the inquest on Sidney Barker, three anonymous letters were sent to the boy's father urging him to sue Maynard for his son's death. All the letters suggested that the 'young lady' who spoke to the inquest would be prepared to help in further proceedings. Did someone know more than had been discovered at the inquest? Also, the poisonings continued. A palpable sense of fear crept through the streets of the seaside town: where and who would the poisoner strike at next? The Police had no leads, and no obvious way of protecting the local population. They decided to make a public appeal. Brighton's chief constable placed an advertisement in the local newspaper offering a reward for any information which led to the arrest of the poisoner.

That action became part of the endgame. Another element was the imminent departure of the Beards from Brighton to a new life in Scotland. The intrigue culminated on Thursday 10th August 1871, when six prominent local men and women, including Mrs. Emily Beard, received parcels of poisoned fruits and cakes, couriered on a train to Brighton from Victoria Station. This time, two of

Mrs. Beard's servants had been invited to taste her gift; they had duly eaten a poisoned plum cake and fallen ill. Mrs. Beard's household was not alone: one of the Beard's neighbours had also been poisoned, along with the editor of the local newspaper. And, once again, Christiana Edmunds had received one of the poisoner's parcels. When the Police arrived to remove her parcel, she told them that that she feared for her safety, as it seemed impossible that the culprit could ever be found. 'How very strange', she said, 'I feel certain that you'll never find it out'. After she had shut the door on the local boys in blue, she took up her pen and paper, and wrote her latest letter to Dr Beard, drawing much attention both to Mrs. Beard's near miss, and to the Barker inquest earlier in the summer.

Christiana was taunting the Police, and she was taunting Dr Beard; in fact, she was taunting everyone. Did she want to be caught? If so, she had sown the seeds of her own capture. It was after he received that latest letter that Dr Beard decided to go to the Police and voice his suspicion that Christiana Edmunds might have something to do with it all. He handed over the large cache of letters which she had continued to write to him, even after her banishment from his presence. That he had kept these letters, secretly, meant that they were potentially incriminating to him as well; but he concluded that the seriousness of the situation required him to face his own, social judgement. The Brighton Police decided to test his theory. They wrote to Edmunds about the Barker case, and received a reply in the same hand as the doctor's correspondence. They decided that the matter warranted further investigations.

Christiana was arrested a week after that last batch of poisoned parcels arrived. Immediately, the Police began to ask around about Miss Edmunds and what she did, and suddenly, many small and unconnected incidents began to make sense. It did not take long to discover that she had left Brighton on Tuesday 8th August to spend two days in Margate, attending to family business. Further enquiries indicated that she had then caught the train to London, before returning to Brighton from Victoria on the Thursday in question. She was on the same train that carried the poisoned post, and had been placed at the scene of the crime. However, what exactly was the crime? The Police worked forwards from Dr Beard's letters. They concluded that the motive must be sex: Christiana was demonstrably in love with Dr Beard, and had decided that her only hope at union lay in the removal of Mrs. Beard from this mortal coil. Edmunds was charged with attempted murder.

This set the scene for her committal hearing, which began at the Brighton Police Court one week after her arrest, on 24th August 1871. Christiana appeared decked in black: a long silk dress, a lace shawl, and a veiled bonnet. Over the course of three hearings over the next fortnight, many witnesses provided pieces in the jigsaw. Dr Beard testified to the events of September 1870, when his wife had fallen sick after eating chocolates. A boy called Adam May testified that he would run errands for Edmunds, taking forged prescriptions to druggists to obtain poisons. He would also purchase sweets and chocolates for her from Maynard's. A chemist called Isaac Garrett testified that he had known Edmunds as 'Mrs. Wood' for four years, and that in March 1871 and two subsequent occasions he had supplied her with strychnine. She had said she wanted to poison some local cats which had become a nuisance. Garrett said that a local milliner called Mrs. Stone had vouched for Edmunds's good character. There were others who were called to the stand, too, placing Edmunds at the scene of other poisoning events, hitherto unknown.

It quickly became apparent that enough evidence existed to charge Edmunds with additional offences. Arsenic had been found in the last batch of parcels, and Edmunds was also known to have purchased arsenic as well as strychnine. Secondly, those who had received the recent poisoned gifts all appeared to know the Beards or have some knowledge of the poisoning case. Most significantly, the name of Maynard's kept returning. It was Christiana who had drawn attention to herself and to Maynard's at the time of the inquest into Sidney Barker's death, when she had provided evidence of her own poisoning. Now, a handwriting expert concluded that the addresses appended to the parcels, the signatures of 'Mrs. Wood' in Mr. Garrett's books, and even the notes handwritten to Sidney Barker's father, were all by the same author as that August letter to Dr Beard. The handwriting was a direct match. That author had also been a regular customer at the sweet shop, placing herself at the centre of all that had gone on in Brighton that summer. The direction of the prosecution changed, probably to Dr Beard's great relief. The case was no longer about his wife, and his relationship with Christiana. On 7th September,

Edmunds was charged with the murder of Sidney Barker, and it was this new charge on which she would stand indicted.

The story now suggested by the prosecution was that after Christiana's failed attempt to poison Emily Beard in September 1870, her subsequent poisoning spree had been occasioned by a wish to blame Maynard's for the whole affair. The suggestion was that by casting guilt elsewhere, Christiana believed she could reassure Charles that he had no grounds to banish her. The truth was that no one was really sure what she had hoped to achieve. An alternative argument doing the rounds was that Christiana had taken to experimenting in preparation for a renewed attempt to kill the obstacle to her own, personal happiness. Throughout the spring and summer of 1871, these experiments had been meted out allegedly on animals and innocent passers-by, with different dosages of poison being trialed and the results noted. Whatever, it was all sensational stuff, and while some of these ideas were purely supposition, the notion of Edmunds's unrequited love driving her to murder was one all too eagerly consumed by the press.

The case was scheduled to be heard at the Lewes Assizes, close to Brighton, until it was felt impossible to find a jury who would not be prejudiced by what they had read in the newspapers. Instead, Edmunds was taken by train to Newgate Prison in London, and her case was heard at the Old Bailey on the 15th and 16th January 1872. She was placed on trial for the murder of Sidney Barker.

The circumstances of the case had set tongues wagging all over the metropolis, and it was not surprising to find the court room full of journalists and other onlookers. Christiana did not disappoint them, appearing once more before the court resplendent in black, this time of velvet with a fur trim. She was bareheaded, and though her age was stated to be thirty-five, for the first time her audience could see that she might be older than those stated years. Her black hair was parted centrally and plaited, so that it was drawn back and down the back of her head. The Times reporter was rather uncomplimentary, suggesting that she had a 'long and cruel' chin, her lower jaw 'massive, and animal in its development'. Despite that, he was prepared to concede that 'the profile is irregular, but not unpleasing', and that there was 'considerable character in its upper features'. Her lips occasionally pressed together in a look of 'comeliness' that turned to 'absolute grimness'. The portrait was painted: a woman who thought herself more than she was, an amatory, predatory woman. It is this caricature that has stayed with her.

She took copious notes of proceedings, her dark eyes flashing up and down as she dipped her pen into the inkwell. The evidence from the earlier hearings was repeated, of poisons purchased and of love gone badly. There were more witnesses by now; various people had come forward to say that Edmunds sent boys to buy sweets for her from Maynard's shop. Shortly after, she would return the sweets, indicating that the wrong ones had been purchased in the first place. These sweets would then be returned to their jar for resale, and alternatives purchased in their stead. There were also witnesses who had seen her leave bags of Maynard's sweets lying around in other shops and public places. Gradually, the events of the last eighteen months came to light.

Her barrister set up the defense of insanity. Several well-known authorities testified on her behalf. Dr William Wood argued that she satisfied the principal MacNaughten Rule – she could not distinguish right from wrong. He had worked previously at Bethlem, and now ran private asylums in London. He was also a regular expert witness in insanity cases. Drs Charles Lockhart Robertson and Henry Maudsley, the famous psychologist, argued that Edmunds belonged to the 'morally defective' group of lunatics – a Victorian precursor to the later term of psychopath. Robertson was a friend of Maudsley's, and the Superintendent of the Sussex County Asylum. He was particularly interested in women's mental health, and had pioneered the use of Turkish baths to calm female patients. Between the three of them they offered a heavy tilt towards a verdict of not guilty, but insane.

Then Edmunds's mother took the stand to deliver a long tale of family madness, which had eventually trapped her surviving daughter. Edmunds, for the only time in court, reacted to proceedings. Contemplating her mother laying bare the family soul, she cried out: 'This is more than I can bear'. In the end, it was futile testimony anyway. As her counsel moved on, Christiana's defense unraveled. There was evidence of hereditary insanity, to be sure, but there was nothing else to offer to back up the opinions of the medical men. There was nothing obviously insane about Edmunds's own life. Any sympathy the court had drifted away from her. When the jury was asked to deliver their verdict, they found Christiana Edmunds guilty of murder, and did not recommend mercy.

The defendant remained in the dock to hear her fate. Neatly dressed, she was still wearing her black velvet cloak with its fur trim. She had added a pair of black gloves to her courtroom attire, and her hair was now arranged 'coquettishly'. Before sentence was passed, she asked to be tried on the original charge too, of attempting to murder Emily Beard, so that she might be able to describe the nature of her relationship with Dr Beard. If she was to go down, she surmised, and then he would go down beside her. It was, of course, too late for that.

Edmunds faced the gallows alone. Her immediate response was fittingly dramatic: she claimed that she was pregnant. It was a legal tradition that a pregnant woman could not be hanged until after she had given birth. A great murmur erupted around the court: so the business of sentencing was not done yet. Immediately, the court officials began to cry out for women of a certain age to make themselves known to them. A jury of matrons was duly empanelled from amongst the spectators in the room, and retired to examine Edmunds in an ante room. A doctor was summoned. The court adjourned until an hour later, when both Edmunds and this latest jury returned to the room. Asked for their verdict, they declared that Edmunds was not pregnant. The law would take its course.

She was returned to Lewes Prison to suffer the extreme penalty of the English legal system. But the medical evidence presented at her trial had not gone unnoticed, and there was popular sentiment locally towards sparing Edmunds's life. On 23rd January 1872, Dr William Orange, by now Broadmoor's Medical Superintendent, visited her together with Sir William Gull from Guy's Hospital at the Home Office's request. Their report summarised her case as follows: 'This woman appears to have had a tranquil, easy and indifferent childhood and womanhood up to a period of about three years ago…The acts were the fruit of a weak and disordered intellect with confused and perverted feelings of a most marked insane character…The crime of murder she seems incapable of realising as having been committed by her though she fully admits the purchasing and distributing the poisons as set forth in the several counts against her. On the contrary she even justifies her conduct'. They declared her to be insane, and after some consideration the Home Secretary, Henry Bruce, respited her sentence to one of Her Majesty's Pleasure.

This was quite an unusual decision, overturning as it did the verdict of a jury. It was not uncommon to have the death sentence commuted to life imprisonment, and there were other Broadmoor murderers who had been transferred with such a tariff. Their guilt, however, remained. Christiana had been absolved from hers by two professionals, contrary to the result in the courtroom. The Times bemoaned this unsatisfactory situation in a leader piece on 25th January, even if it did agree that the outcome had been the right one. It wondered aloud on the wisdom of politicians permitting a jury to give 'a solemn verdict which they know will be afterwards reversed'. The decision was unpopular back in Brighton too: the Home Secretary had effectively saddled the ratepayers with Christiana's upkeep from now on, creating another large bill to pay. Certainly her case had been a big ticket item, making full use of venues, discourse and precedent. Perhaps the attention was thrilling, though the fact that a verdict could be legally correct yet medically unsound was a conclusion of little importance to Christiana. She had achieved a more basic ambition. Gull and Orange had given her back her life, and she was therefore transferred to Broadmoor as a pleasure patient on 5th July 1872.

On her arrival at the Asylum, she was forty-three years-old. She was wearing make up on her rouged cheeks, a wig ('a large amount of false hair') and had false teeth. 'She is very vain', wrote Dr Orange at the time. The surgeon at Lewes Prison who signed her transfer documents had obviously done so reluctantly. He was most unimpressed with the diagnosis of insanity, writing that after ten months of supervision he could not be satisfied either that Edmunds was insane, or that she was not responsible for her actions. He did, however, say that she was of a delicate constitution, and prone to being hysterical.

Dr Orange was nevertheless convinced that he had made the correct diagnosis. Edmunds's behavior in his charge did not conform to social norms. When her surviving brother died shortly after her admission, she showed no grief, and appeared to be completely unmoved by the loss. She was also deceitful. As soon as she was transferred, she immediately began to try and smuggle in clothes or beauty aids. Her younger sister, Mary, was complicit in this. One letter asked for clothing; another talked about ways to find and apply make-up while in the Asylum. Orange attempted to reason with Mary, insisting that Christiana was able to partake of any comfort that she required. It was to no avail. Mary began to send Christiana gifts too, and it was the gifts that caused great irritation to the matron of Broadmoor's female wing. Inside every parcel was some sort of contraband, hidden within another item. Each one needed time and attention to search. It appeared to be attention-seeking on the part of both of the Edmunds women, and it was more than the matron could bear. The final straw was the receipt of a cushion stuffed with false hair during 1874. The matron complained to Orange that Edmunds was amassing and hoarding hair in her room, and that no further gifts should be allowed. The Superintendent was initially reluctant to interfere with behavior which he saw as self-indulgent, but largely harmless. The matron, however, put her foot down.

Also in 1874, Broadmoor intercepted clandestine correspondence sent to the chaplain at Lewes Prison, with whom Christiana had struck up a bond during her time in custody. Dr Orange noted that he had no objection at all to Edmunds corresponding with the chaplain, but her decision to do so secretly was 'in conformity with her state of mind to prefer mystery and concealment'. Presumably the chaplain was intended to become a Dr Beard substitute. Still, Christiana's webs of intrigue continued. In 1875 her room was twice searched and various concealed articles were recovered on each occasion. Dr Orange wrote that 'she deceives for the pure love of deception'.

Edmunds was a patient who required micro-management. She was a bundle of contradictions. Generally quiet and biddable, she joined the ranks of the more trusted patients in the original female Block. She had access to the Terrace and the gardens, and probably delighted in causing mischief through playing croquet and other games with her fellow patients. For she was certainly disruptive, as a note of 1876 indicates: 'her delight and amusement seem[s] to be in practicing the art of ingeniously tormenting several of the more irritable patients so that she could always complain of their language to her whilst it was difficult to bring any overt act home to herself'. The same note suggests that her room is still being regularly searched, and that when her mother visited, she would omit her make up and try to look as desperate as possible.

The subject of Christiana's make-up appears often in her notes. She was evidently perceived by the male doctors as Broadmoor's painted lady, and as a creature motivated by romantic desire. They were the sole males in regular contact with her, and she appears to have been determined to maximize their attention to her. A note made in 1877 by David Nicolson, as Edmunds approached the age of fifty, related her daily life as one of embroidery and etching; but also maintained that she 'affects a youthful appearance' and that 'her manner and expression evidently lies towards sexual and amatory ideas'. It seems certain that at the annual Christmas dance for female patients, no doctor or male attendant could escape a dance with Christiana.

Her life at Broadmoor continued in this vein for another thirty years. She presented no danger to any staff or patients, and unlike some patients she showed no obvious signs of insanity. Many times her notes described her as being obsessed with her personal appearance. She won the battle to wear her own clothes eventually. We know this because she sent out a parcel of them to a Wokingham lady for repair in 1887, and the parcel was sent back to the Broadmoor steward, who made a fuss because he was not expecting it. Otherwise, she became less disruptive. She sewed, she painted, she made herself up and demanded acknowledgement from the male staff when she met them; she was quiet, she was well-behaved, and she showed no remorse for her crimes. And in doing all these things, she grew into an old woman.

Perhaps if she had been one of the Broadmoor women who had acted while suffering from post-natal depression, she might have been discharged. But there was no clamour for that, nor any regular petitions to the Home Office, letters in the newspapers or campaigning friends to ask questions on her behalf. Dr Orange even noted in 1884 that he did not actually have any paperwork authorising her detention, because the Treasury Solicitor had lost it all. It never seems to have crossed anyone's mind that she might be discharged to rejoin society. As the years went by, her remaining family died, and she was left alone at Broadmoor.

Gradually her own health weakened. In 1900, she was bedridden for a while with flu. By 1901 her sight was fading badly, and she could barely see out of her right eye. She rallied in time to attend the Asylum's annual ball in 1902, but her mobility decreased, and by 1906 she could hardly walk to go anywhere. As she entered the last year of her life, a final Christmas ball approached. Laid up in the infirmary, and closely observed by the medical staff, a snippet of conversation between her and another patient was entered into her case notes:

Edmunds: How am I looking?

A: Fairly well.

Edmunds: Are my eyebrows alright?

A: Yes.

Edmunds: I think I am improving. I hope I shall be better in a fortnight. If so, I shall astonish them; I shall get up and dance – I was a Venus before and I shall be a Venus again!

She died nine months later on 19th September 1907, aged 78. The cause of death was given as senile decay, or old age.

Edmunds had a lasting effect on many of the professionals around her. Her case had been notable, and Dr George Blandford used it to illustrate his book Insanity and its Treatment, quoting Dr Orange's original report on Edmunds. In 1892, Blandford was preparing a new edition of his book, and wrote to Nicolson, Orange's successor, asking if he could have an update on how Edmunds had changed during her twenty years at Broadmoor. Dr Nicolson replied that he had seen no change in Edmunds during the fifteen and a half years that he had known her.

Most significantly, hers was apparently the first capital trial witnessed by the great English barrister Sir Edward Marshall Hall. Marshall Hall would later make a name for himself by taking on the defense case in a number of high profile English murder trials, earning himself the title of 'The Great Defender'. Another Brighton resident, he was only thirteen at the time of Edmunds's trial, but it is generally accepted that he joined other spectators at the Brighton Police Court hearings, and perhaps he was captivated by the undoubted sense of legal theatre which surrounded Edmunds and her woman in black persona.

This sense of performance was something that attached itself to Edmunds, and as a result her case has leant itself to dramatization. She was the subject of an ITV Saturday Night Theatre film as part of its Wicked Women season in 1970, where Anna Massey starred as Edmunds. The story has also been broadcast as The Great Chocolate Murders on BBC Radio 4 in 2006, and recently become part of Steve Hennessy's series of Broadmoor plays.

In Brighton, Christiana and the other characters in her story are still well-known and used regularly in written or dramatic works. The facts of the case have become a popular path traveled by those interested in Victorian true crime. The facts have told a story, though still an incomplete one, for Edmunds leaves behind a sense of mystery in terms of her motivation. She is a character who always seems within grasp and then disappears beyond reach. She never denied her actions, nor offered up an explanation of what she was trying to achieve.

She was certainly a slave to adulation, and must have thrived on the publicity that her criminal actions generated. She must also have enjoyed the secrecy attached to the affair on which she embarked with her doctor neighbour. Perhaps her motive was no more than to enjoy all these experiences. It is unclear whether she wanted to have Dr Beard or to ruin him, and there is no firm evidence that she ever sought to correspond with him again after August 1871. It is, though, too neat an ending to conclude simply that all was vanity with her: that this unusual woman can be reduced to a female stereotype, a frustrated spinster whose desires eventually destroyed her. Not enough of her survives in the records to be able to see the true Christiana, and she has left us with only shards of the mirror containing her reflection. The search to discover the Venus of Broadmoor goes on.

As we move towards Broadmoor Hospital the insane criminals were housed in the high-security psychiatric hospital at Crowthorne in the Borough of Bracknell Forest in Berkshire, England. It is the best known of the three high-security psychiatric hospitals in England, the other two being Ashworth and Rampton. Scotland has a similar institution at Carstairs, officially known as The State Hospital but often called Carstairs Hospital, which serves Scotland and Northern Ireland. The Broadmoor complex houses about 260 patients, all of whom are men since the female service closed with most of the women moving to a new service in Southall in September 2007, a few moving to the national high secure service for women at Rampton and a few elsewhere. At any one time there are also approximately 36 patients on trial leave at other units. Most of the patients there suffer from severe mental illness; many also have personality disorders. Most have either been convicted of serious crimes, or been found unfit to plead in a trial for such crimes. The average stay for the total population is about six years, but this figure is skewed by some patients who have stayed for over 30 years; most patients stay for considerably less than six years. The catchment area for the hospital underwent some rationalisation of the London area in the early twenty-first century, and now serves all of the NHS Regions: London, Eastern, South East, South West. One of the therapies available is the arts, and patients are encouraged to participate in the Koestler Awards Scheme. The hospital was previously known as the Broadmoor Criminal Lunatic Asylum; the change of name reflects a change in attitude towards mental illness, criminals, and the word "asylum". The hospital was built to a design by Sir Joshua Jebb, an Officer of the Corps of Royal Engineers, and covers 210,000 square metres (53 acres) within its secure perimeter. It received its first female patients on 27 May 1863, with the first male patients arriving on 27 February 1864. The original building plan of six blocks for men and two for women was completed in 1868. A further male block was built in 1902. Due to overcrowding at Broadmoor, a branch asylum was constructed at Rampton Secure Hospital and opened in 1912. Rampton was closed as a branch asylum at the end of 1919 and reopened as an institution for mental defectives rather than lunatics. During World War I Broadmoor's block 1 was also used as a prisoner-of-war camp, called Crowthorne War Hospital, for mentally ill German soldiers. After the escape and the murder of a local child in 1952 by John Straffen the hospital set up an alarm system, which is activated to alert people in the vicinity, including those in the surrounding towns of Sandhurst, Wokingham, Bracknell and Bagshot, when any potentially dangerous patient escapes. It is based on World War II air-raid sirens, and a two-tone alarm sounds across the whole area in the event of an escape. It is tested every Monday morning at 10 am for two minutes, after which a single tone 'all-clear' is sounded for a further two minutes. All schools in the area must keep procedures designed to ensure that in the event of a Broadmoor escape no child is ever out of the direct supervision of a member of staff. Sirens are located at Sandhurst School, Wellington College, Bracknell Forest council depot and other sites. Following the Peter Fallon QC inquiry into Ashworth Special Hospital, which found, amongst other things, serious concerns about security and abuses that came about from poor management, it was decided to review the security at all three special hospitals. Until this time each special hospital was responsible for maintaining its own security policies. This review was made the personal responsibility of Sir Alan Langlands who at the time was Chief Executive of the National Health Service (England). The report that came out of the review initiated a new partnership to be formed whereby the Department of Health sets out a policy of safety and security directions that all three special hospitals must adhere to. These directions are then updated or modified as needed. This has resulted in upgraded physical security at Broadmoor from approximately category 'C' to category 'B' prison standards. Higher levels of security than this are then placed around certain buildings. New standards have also been formulated to increase procedural security and safety for the staff and other patients; these include procedures and equipment for reducing the amount of contraband smuggled into the hospital. Before the Langlands report, it had been an anathema in modern psychiatry to think of enclosing the mentally ill behind razor wire. As this type of security measure had been seen as unnecessary, it was thought that it would only serve to reinforce the stigma against psychiatric patients if it were to be employed.

Broadmoor the Prison?

Because of the outside appearance of the buildings, especially its high walls and other visible security features, and the inaccurate news reporting it has in the past received, it is occasionally presumed by some members of the general public that Broadmoor Hospital is a prison.[7] Many of its patients are indeed referred to it by the criminal justice system, and its original design brief incorporated an essence of addressing criminality in addition to mental illness; however, the layout inside and the daily routine are designed to assist the therapy practised there rather than to meet the criteria necessary for it to be run along the lines of a prison in its daily functions. Many staff were often members of the Prison Officers Association, as opposed to the health service unions like UNISON. Jimmy Noak, Broadmoor's director of nursing, in response to claims that criminals were being given unfairly pleasant treatment in the facility, commented, "It's not fair, but what is the alternative? If these people committed crimes because they were suffering from an acute mental illness then they should be in hospital." Governance. From its opening until 1948

Broadmoor was managed by a Council of Supervision, appointed by and reporting to the Secretary of State for the Home Department (Home Secretary). Thereafter, the Criminal Justice Act of 1948 transferred ownership of the Hospital to the Department of Health (and the new NHS) and oversight to the Board of Control for Lunacy and Mental Deficiency established under the Mental Deficiency Act 1913. It also renamed the hospital Broadmoor Institution. The Hospital remained under direct control of the Department of Health - a situation which reportedly "combined notional central control with actual neglect"[9] until the establishment of the Special Hospitals Service Authority in 1989, with Charles Kaye as initial Chief Executive. In 1996 the SHSA itself was abolished, being replaced by individual special health authorities in each of the High Secure Hospitals. The Broadmoor Hospital Authority was itself dissolved on 31 March 2001.[10] Then on 1 April 2001 West London Mental Health (NHS) Trust took over the responsibility for this hospital. This Trust reports to the NHS Executive through the London Strategic Health Authority.

The Paddock Centre

A new unit called the Paddock Centre was opened on 12 December 2005 to treat patients with a dangerous severe personality disorder (DSPD). This is a new and much debated diagnosis or label that has two criteria. The first criterion is that the individual be 'dangerous', i.e. they are considered to be or represent a 'Grave and Immediate Danger' to the general public. It has been suggested that the threshold for this criterion be set at a greater than 50% chance of that individual committing serious harm upon another, from which the victim is unlikely to recover. The second DSPD criterion is that the individual suffers from a 'severe personality disorder', meaning that he or she has: A diagnosis of two or more personality disorders that meet the criteria as laid out in the Diagnostic and Statistical Manual of Mental Disorders DSM IV –TR; or A significant score (i.e. 30 or higher) on the Hare Psychopathy Check list – Revised (PCL-R); or A slightly lower score (i.e. 25 to 29) on the Hare Psychopathy Check list and with one or more personality disorders but not including an Antisocial personality disorder diagnosis.

Rather than create a new Mental Health Act, it may now only require the existing laws to be updated in order that people can be assessed for this condition before they have been committed to the forensic services by another route. The DSPD service in the Paddock Centre will be limited to men, as it is not yet scientifically agreed whether any women meet this criterion. Individuals who do meet this criterion will be admitted to the new Paddock unit only as and when sufficient staff have been trained to be able to provide and maintain the right therapeutic programmes and environment. The Paddock Centre is designed to eventually house 72 patients, and is one of four units being set up in England and Wales. The architects were Oxford Architects LLP. As the West London Mental Health NHS Trust already carries out research, the Trust hopes that Broadmoor will become a centre of learning for this new type of therapy. The ultimate aim of this work is to reduce the cost to society compared to that which would accrue if no treatment was provided. During the second half of the century public asylums became dominant as private madhouses declined in number. The body of legislation on lunacy was consolidated in the Lunacy Act of 1890, which remained the basic legislation until 1959, and a further act of 1911 authorised the appointment of two further commissioners because of increased work. The latter was one of the minor recommendations of the report in 1908 of the Royal Commission on the Care and Control of the Feeble-Minded, whose appointment in 1904 followed years of public pressure for recognition of mental deficiency as a condition distinct from lunacy. Its main recommendations were embodied in the Mental Deficiency Act 1913. The Mental Deficiency Act 1913 merged the Commissioners in Lunacy into a new Board of Control, with enlarged powers including the supervision of the mental deficiency service established by the act and run by local authorities through mental deficiency committees. The board was subject to the general superintendence of the home secretary. In May 1920 most of the functions of the Home Office in relation to the Board of Control passed to the new Ministry of Health. The Home Office retained responsibility for the detention of criminal lunatics and for the administration of Broadmoor until it was handed over to the board in 1948. The Lord Chancellor remained responsible for regulations under the Lunacy and Mental Treatment Acts. Pressure for reform resulted in 1922 in the appointment of a departmental committee to enquire into the administration of mental hospitals, and in 1924 of a Royal Commission on Lunacy and Mental Disorder, whose report in 1926 recommended that the Board of Control be given some of the executive functions of the Minister of Health. Under the resulting Mental Treatment Act 1930 the board acquired duties concerning the reception, care, treatment and discharge of patients in mental hospitals and facilities for the treatment of such cases. In July 1947 the statutory functions of the Board of Control, with the exception of its quasi-judicial powers relating to the protection rather than the medical treatment of patients, and many of its staff, were transferred to the Ministry of Health. The board remained responsible until 1960 for the special hospitals for the criminally insane at Rampton and Moss Side which had been transferred to it in 1920. Also in 1947 the Office of the Master in Lunacy was renamed the Court of Protection. On 27th May 1863, three coaches pulled up at the gates of a recently-built national institution, which had been set amongst the tall, dense pines of Bracknell Forest. Inside these three coaches were eight women and their escorts from Bethlem Hospital in London, the ancient hospital for the treatment of the insane. It was now early afternoon, and that morning, the little party had left the Bethlem buildings in Southwark, boarded a train at Waterloo and been taken by steam through the capital's suburbs and out to the little market town of Wokingham in Berkshire. Their destination was Broadmoor, England's first Criminal Lunatic Asylum. At half past twelve, they had alighted from the train at Wokingham's simple railway station and found the three coaches waiting for them: a larger one, grandly-titled the Broadmoor Omnibus, together with two smaller vehicles. These carriages would take them on the last leg of their journey. The eight women and their accompanying paperwork were loaded into the seats, before the steps were removed and the horses started. Then the wheels of the coaches spun down winding earthy lanes and finally up a gentle incline as the passengers were driven the five miles to Crowthorne. Broadmoor's first patients had arrived. Who were these women? As befitted a group thrown together without friendship, they had different backgrounds. One was a petty thief, for example, while another had stabbed her husband when they were out poaching. Then there were the other six, who had all shared a single life event. They had killed or wounded their own children: either strangling them, drowning them, or cutting their throats with a razor. It was one of this last group who was the first patient to be listed in the new Asylum's admissions register. Her name was Mary Ann Parr. She was about thirty-five years of age, and a labourer from Nottingham. She had lived in poverty all her life, almost certainly suffered from congenital syphilis, and had what we would now call learning disabilities. Mary might have been just another member of the industrial poor, except that when she was twenty-five years old, she had given birth to an illegitimate child and then suffocated it against her breast.

She had been convicted of murder and sentenced to death, but her sentence was commuted first to transportation for life, and then, after a medical examination, to treatment instead in Bethlem. When Mary Ann Parr arrived at Broadmoor, as with every patient who would come after her, her details were first recorded from the forms that had accompanied her, and then she underwent a medical examination and an interview with one of the doctors. All the while, notes were taken, and these notes were then written up into a large case book, and added to over the years. This is an extract from the notes made about Mary Ann Parr on admission: 'A woman of weak intellect, complains of pains in the forehead, short stature, cataract of the left and right eyes – can see a little with the left eye only. Teeth irregular and notched…Of very irritable temper.' Mary Ann Parr and the other new patients were given the best treatment that was available at the time. This was rather different to how we might understand mental health treatment today. There were no drug therapies available for the mentally ill during Victorian times, or psychiatric analysis. Instead, Her Majesty's lunatics were subject to a regime known as 'moral treatment'. This was a recognisable Victorian concept. Mary was given a regular daily routine of exercise and occupation (which for her meant working in the laundry); regular meals of fairly bland food; and plenty of fresh air. She was also given relief from her poor and harsh surroundings. Her quality of life was probably significantly better than that she had enjoyed outside: she had a roof over her head, and she did not have to worry about food or money. This removal of a patient from their usual society was another aspect of Victorian treatment. By giving a patient refuge in the Asylum, the Victorians believed they would be able to neuter the immediate causes of insanity in their day-to-day life, leading to beneficial results. It was a recognition that community living could create problems as well as solutions. Mary Ann Parr was a reasonably typical recipient of this treatment regime, in that she experienced it for the next thirty-seven years, until she died in 1900, aged seventy-one, from kidney disease. Many patients spent decades on site, and became institutionalized in the process.

The Turn of the 19TH Centaury

By the time that Queen Victoria finally relinquished her grip on the British throne, Broadmoor had become a recognised part of the medical, judicial and social landscape. It was a bigger place than it had been in 1863, though the wide range of needs for which it catered remained roughly the same.

What remained of the Victorian Asylum in 1901 still remains today. This is not just the bricks and mortar, but the records from that time, and it is these records that have been used to draw together the stories that follow. This short collection does not pretend to be a complete history of the hospital during the Victorian period, and rather is meant to encourage other researchers to focus on particular aspects of that time. One of the incredible features of the archive is that there is something for everyone. The stories are true, the people are real, and the history is there to be discovered. So enjoy your brief tour round Victorian Broadmoor.

Edward Oxford

Edward Oxford was a young man who became famous, or more properly infamous, in Victorian Britain. It was a state that he said he had aimed for, and to that end took aim at Her Majesty Queen Victoria in a probably not very serious assassination attempt. His actions led both to his notoriety and to over twenty-five year's detention at Her Majesty's Pleasure.

He was born in Birmingham on 19th April 1822, the third of seven children to Edward and Hannah Oxford. His childhood was spent in both Birmingham and Lambeth. Although his father died when Oxford was seven, his mother was always able to work, and he was sent to school in both places. Oxford and his mother remained close, despite their occasional parting due to her working habits.

After Oxford completed his schooling he took bar work, first from his aunt in Hounslow and then later at other public houses. By the age of eighteen he had grown up to be a pale youth, with brown eyes and auburn hair, around five foot six inches tall. At the start of 1840, he was working as a pot boy (barman) in The Hog in the Pound along Oxford Street in London, and living with his mother and sister in lodgings in Camberwell. He quit this job at the start of May 1840 without having further work to go to.

A week after he quit his job, his mother returned to Birmingham on a regular trip to see family, and Oxford was largely left to his own devices. Some five weeks later, on the late spring evening of 10th June 1840, he took up a position on a footpath at Constitution Hill, near Buckingham Palace. It was 6pm. He waited for the young Queen Victoria and Prince Albert to be driven out from the Palace in an open carriage, and when they drew level with him, he fired two shots in succession from separate pistols at the Queen. She was four months pregnant at the time with her first child, Victoria, the Princess Royal.

Immediately, various members of the public seized Oxford and disarmed him. Oxford was quite open about what he had done, exclaiming 'It was I; it was me that did it.' What was not clear was exactly what he had done: he had certainly fired two pistols at their Majesties, but whether those pistols could have harmed anyone was never resolved. No bullets were ever found, and the Crown was unable to prove that the pistols were armed when Oxford discharged them. Once sentenced, Oxford always maintained that the pistols contained only gunpowder.

Oxford was arrested and charged with treason. After his arrest, his lodgings were searched and a box found, which amongst other fragments of his life contained the intricate rules he had constructed of a fictitious military society called Young England, complete with imaginary officers and correspondence. Members were to be armed with a brace of pistols, a sword, a rifle and a dagger.

Inevitably, his trial attracted much attention, and it was postponed until a thorough investigation had been made into both his background and his possible motives. Insanity was used as his defense. On Thursday 9th July, the Old Bailey was packed with those citizens fortunate enough to have obtained a ticket for admission. Oxford appeared largely oblivious to proceedings. The prosecution presented a large amount of witness evidence before various family members and friends testified that Oxford had always seemed of unsound mind and that both his grandfather and father had exhibited signs of mental illness and were alcoholics. This was important to the Victorians, for whom both drink and hereditary influence were strong causal factors for insanity. His mother, in particular, told a sorry tale of domestic violence and intimidating behavior from Oxford's father, which was rich in detail and must have had quite an impact at the trial. Oxford himself, she said, had always cried without apparent cause, and been prone to fits of hysterical laughter. When she had worked in a shop he would annoy the customers by laughing or making strange noises, and had been obsessed with firearms since he was a child.

Oxford's principal medical witnesses were Dr Thomas Hodgkin, who considered that he had a 'lesion of the will' – that he could not control his impulses – and Dr John Conolly, Head of the Hanwell Lunatic Asylum (now St Bernard's Hospital, Ealing), who believed that Oxford had suffered a disease of the brain, as evidenced by the shape of his head. Conolly had asked Oxford why he shot at the Queen, and Oxford replied 'Oh, I may as well shoot at her as any body else.' The defense called other medics too - Dr William Dingle Chowne agreed that Oxford could not control his will; while Dr James Fernandez Clarke thought Oxford was a hysterical imbecile. All agreed that Oxford was of unsound mind.

These were significant names in Victorian medicine. Conolly was the man who had destroyed every form of restraint used at Hanwell and promoted a new 'moral' regime of mental health care through routine and responsibility. At the time of Oxford's trial the controversy surrounding his new ideas was in full swing. Clarke was an acclaimed medical author and a major contributor to The Lancet, while Hodgkin was an eminent pathologist who gave his name to Hodgkin's disease. Chowne was a respected manager at Charing Cross Hospital and a leading advocate of sanitary reform.

The next day, the jury returned to acquit Oxford on the grounds of insanity. He received the sentence of all such lunatics – to be detained until Her Majesty's pleasure be known, effectively an indefinite sentence, and one which gave rise to the Broadmoor term of 'pleasure men'.

Within weeks, Oxford had been removed to the State Criminal Lunatic Asylum at Bethlem, then in Southwark, to begin his sentence. Some notes from Bethlem were copied up into his Broadmoor case notes. The entry for 16th February 1854 stated that 'No note has ever been made of this case, and no record kept of the state of his mind at the time of his admission, but from the statements of the attendants and those associated with him he appears to have conducted himself with great propriety at all times.' Indeed, he seems to have become a model patient, industrious and studious. He spent much time drawing, reading and in study, learnt French, German and Italian to a standard of virtual fluency, while obtaining some knowledge of Spanish, Greek and Latin, as well as learning the violin. The Bethlem doctors also reported that he could play draughts and chess better than any other patient. He also became a painter and decorator, and was gainfully employed within the Hospital. Of his crime, the notes stated that 'He now laments the act which probably originated in a feeling of excess vanity and a desire to become notorious if he could not be celebrated.'

Presumably his positive influence on the ward was missed by the Bethlem authorities when he was moved to Broadmoor on 30th April 1864, even if in general the London hospital was happy to be rid of the criminal lunatic class.

Oxford's health on arrival in Broadmoor was stated to be good, though he suffered from constipation and some oedema (swelling) in his lower legs. By this date he was forty-two years old, and had been confined for nearly twenty-four years.

His notes on arrival in Broadmoor record: 'A well conducted industrious man apparently sane, has been rather out of health since last Christmas and has suffered from urethritis since his admission here – this he attributes to his having taken various unusual things to drink just before leaving Bethlem. He is now in better general health. He states that he fired a pistol charged with powder only at the Queen on June 10th 1840. That he did it under the impression that he should thereby become a noted person and that he had not the smallest intention of injuring Her Majesty.'

He carried on his diligent application to hard work at Broadmoor, working daily as a wood grainer and a painter and being very well-behaved. It was increasingly obvious that Oxford no longer posed a risk to anyone, and that he was also completely sane. Sir William Hayter, the Chair of Broadmoor's scrutiny body, the Council of Supervision, wrote to Home Secretary Sir George Grey in November 1864 stating that Oxford was of sound mind. Not only did John Meyer, Broadmoor's Medical Superintendent, testify to this, but also Charles Hood, a member of the Council and Oxford's previous physician at Bethlem. Hood said that Oxford had been sane since at least 1854, when the patient was first in his care. Hayter suggested that Oxford was perfectly capable of being allowed to make his own way in the world.

Grey ignored the request. He had been Judge Advocate General in the Government in 1840, and perhaps he was uncomfortable with allowing the discharge of a case in which he probably had an interest. Instead, Oxford stayed on in the Asylum until September 1867, when new Home Secretary Gathorne Hardy began the process of agreeing to Oxford's discharge when he asked Hayter to provide an up-to-date report on Oxford's mental condition. Subsequently, Hardy offered Oxford release on condition that he went overseas to one of the colonies, and never returned to the United Kingdom. Oxford indicated that he was willing to accept the terms.

Meyer proposed that he arrange a passage to Australia for Oxford. Before Oxford's discharge, the patient was visited by twelve officers from the Metropolitan Police, who took notes about his appearance and photographed him, should he attempt to return. It was made clear to him that if he ever set foot again on the British Isles, he would be locked up for good. Sadly, no copy of the photograph survives in the Broadmoor archives.

The warrant for Oxford's release arrived at Broadmoor towards the end of October. His passage was arranged for a month later. Accompanied by Charles Phelps, the Steward at Broadmoor, Oxford traveled to Plymouth on 26th November 1867. The next day he boarded HMS Suffolk for Melbourne. He remained on board for several days, waiting as the ship was detained in port, until she finally left on 3rd December. Phelps was made to sign an affidavit that 'To the best of my knowledge and belief Oxford was on board when she sailed.'

Oxford certainly sailed to Australia, though the rest of his life is less well documented. In the Broadmoor archive, the only subsequent intelligence about Oxford comes from a letter from George Haydon, one time Steward at Bethlem, to Dr David Nicolson at Broadmoor in 1883. Haydon quoted an article from The Age, a Melbourne newspaper, of which he had been made aware. The article, included with the letter, is about a man called John Oxford, and is dated 4th May 1880. John Oxford was named as the man who shot at the Queen many years ago, and had subsequently been a patient in an asylum before he was discharged to Australia. He had recently been convicted of stealing a shirt and spent a week in jail. Upon his release, the prison governor had asked the police to keep an eye on him, 'in consequence of the old man's eccentric conduct'. To that end the police had arrested Oxford for vagrancy, and the article reported that he was up before the bench again. He was remanded for further medical examination. Haydon's update ended there.

Sources indicate that there is further correspondence from Haydon elsewhere to suggest that Oxford later changed his name to John Freeman, and published a book called Lights and Shadows of Melbourne Life in 1888. Certainly the book exists, but there is nothing in the Broadmoor archive which confirms that he was its author. These other sources quote Haydon as reporting that Oxford was a house painter by trade (carrying on the skills he learnt in hospital) and had married at some point before 1888. Oxford's suggested date of death is 1900.

Queen Victoria suffered several other assassination attempts during her reign, mostly from subjects who, if not legally insane, were certainly considered by the general population to be mad. One of those was another Broadmoor patient, Roderick MacLean, who shot at her at Windsor Railway Station on 2nd March 1882. MacLean was sent to Broadmoor after his trial, but unlike Oxford he did not recover, and remained there until his death in 1921. It was MacLean's case that resulted in a change in sentence for those found to be criminal lunatics, from the traditional 'not guilty by reason of insanity', to the more condemnatory 'guilty, but insane'. The motivation for the law change is always leveled at the Queen's response to MacLean's not guilty verdict: 'Insane he may have been, but not guilty he most certainly was not, as I saw him fire the pistol myself.' This is not entirely true: the Queen did not see MacLean shoot, though she did hear the report of his pistol. However, her displeasure at MacLean's innocence was real, and she pressurized Prime Minister Gladstone to change the law. It is unclear exactly what Victoria hoped to achieve by this, though she alluded to the view that if Edward Oxford had been hanged all those years ago, it might have deterred those potential regicides who came after him. Forty years of being shot at had not mellowed Her Majesty.

Discharges

The discharge rate on the male side was around one in ten, and even greater on the female side, with slightly more than one in three patients being discharged. This was, in part, due to the patient make up. While the 'pleasure' men and women's fate lay ultimately with the Home Secretary of the day, a significant proportion of patients arrived from the prison system with a fixed sentence. Once that sentence was complete, they were usually discharged to a local asylum for care. The fact of Broadmoor's opening does not explain the fact of Broadmoor's creation. Every story has a beginning, and in Broadmoor's case this is usually traced back to a spring day in 1800. It was on the evening of 15th May that year that King George III chose to attend the Theatre Royal in Drury Lane, London, only to feel the whistle of two shots pass near him before he had taken his seat in the royal box. The assailant was a member of the audience. James Hadfield was a young father from London convinced that he needed to secure his own death at the hands of the state. By suffering the same fate as Christ, Hadfield believed that his personal sacrifice would benefit all mankind by ushering in the Second Coming, and the Day of Judgement. This was a fact that would emerge later. For now, Hadfield was restrained in the orchestra pit of the Theatre as pandemonium raged around him. It was clear that Hadfield was mad. Legally, though, he presented a problem. While he might be found not guilty by reason of insanity, this verdict was reserved historically for those described as 'brutes' or 'infants'. The usual result was a discharge, sometimes to Bethlem, London's historic hospital for the mad, more often to family or the local community for care, but certainly with no further oversight from the state. Such a discharge would have been extremely risky in Hadfield's case, as it seemed entirely plausible that if let go, he might try something similar again. Besides, Hadfield was neither brute nor infant. He was married, in regular employment in the silver trade, a war hero, as well as a family man. His case bore some similarities to those of two previous assailants on the Royal person, Margaret Nicholson and John Frith, neither of which had been resolved satisfactorily from a legal point of view. The memory of Nicholson and Frith would have been fresh in the minds of the lawyers brought in to deal with Hadfield. Now, the law was presented with another opportunity to find a way of managing the dangerous lunatic, and the English legal system was helped out of its difficulties to no small extent by the success of Hadfield's lawyer, Thomas Erskine. Today we would consider a defense lawyer to be an automatic right, but they were a bit of an oddity in court until the 1830s, and it was only because Hadfield had been charged with treason that the ancient statutes granted him a right to counsel. Erskine took advantage of this position to argue a revolutionary defense: that the law actually allowed for partial insanity; that is, it included recognition of those people who suffered from bouts of periodic mental illness, and otherwise enjoyed periods of lucidity. Hadfield was such a person. He was diligent and rational when he was not in a religious frenzy. He was found not guilty, and remanded to Bethlem while Parliament regulated the judge-made law. The result was the passing of the Criminal Lunatics Act 1800. This Act gave Hadfield his new status, and the law the power to detain him until 'His Majesty's pleasure be known', the legal form for an indefinite sentence. Duly sentenced, and despite a brief escape from Newgate Prison, Hadfield remained a guest of His Majesty until his death in 1841. Of course, with the new sentence there quickly came further Hadfields, all similarly afflicted and all requiring some form of secure accommodation. As luck would have it, Bethlem had outgrown its city space and was on the verge of moving to larger premises, so the Government negotiated the first dedicated space for criminal lunatics when the new Bethlem opened in St George's Fields in 1816. Two new wings were built as what became known as the State Criminal Lunatic Asylum. It was an opportunist move, rather than a long-term one. When space at Bethlem reached capacity a few decades later, further space was purchased at Fisherton House Asylum in Salisbury, though this also only bought a little more time. As the national population mushroomed during the nineteenth century, so too did the small subset that comprised the criminal lunatics.

So the journey of madness ended at Broadmoor. The treatment for 'the mad' had long been a form of brutal dominance, with restraint, bloodletting, purges and vomits being used unquestioningly The Enlightenment of the 1800s saw a change in attitude towards 'moral management' of patients. This built upon the beliefs of William Battie, but 'madhouses' did provide the opportunity for medical professionals to gain close up experience of patients. From this emerged the model of 'madness' as a condition of the mind rather than the body - the birth of psychiatry. Physician to St Luke's Asylum in London, that 'management did more than medicine.' John Connolly, physician to Hanwell Asylum, Middlesex, was one of the key figures to pioneer the system of non-restraint in Britain in the 1840s and 50s: 'Restraint and neglect are synonymous. They are a substitute for the thousand attentions needed by a disturbed patient.' The medical staff adopted new practices at Bethlem from 1815-1851, in particular advocating the therapeutic value of work for patients. However, although the use of restraint was modified, Bethlem was reluctant to dismiss it entirely and often used traditional therapies alongside new ideas. In this Bethlem was not alone – in the mid-1840s only five county asylums had abandoned mechanical restraint. However by 1854 this number had risen to twenty seven. The final public investigation into reports of patient neglect at Bethlem took place in 1851. It led to the hospital being included in the national system of investigation into asylums and hospitals for the mentally ill. A period of reform followed which led to more enlightened treatment under new management. The ability to change was arguably helped by the fact that there were now, for the first time in more than 120 years, no Monros at Bethlem.

Bibliography

James Tilly Matthews: A patient in Bethlem Royal Hospital from 1797 to 1814

Owen Bowcott. 'Bedlam' exhibition traces the meandering history of mental health

Bedlam: London and its Mad – Catharine Arnold (2008)

'Houses of Military Orders: St Mary of Bethlehem',

A History of the County of London: Volume 1:

London within the Bars, Westminster and Southwark (1909),

Bahat, Dan (1986). "Does the Holy Sepulchre church mark the burial of Jesus?". Biblical Archaeology Review

Bokenkotter, Thomas (2004). A Concise History of the Catholic Church. Doubleday. pp. 155.

Coüasnon, Charles (1974). The Church of the Holy Sepulchre in Jerusalem. London: Oxford University Press

Fergusson, James (1865). A History of Architecture in All Countries. London: J. Murray.

Foakes-Jackson, Frederick John (1921). An Introduction to the History of Christianity, A. D. 590-1314. London: Macmillan. http://www.archive.org/details/introductiontohi00foak.

Gold, Dore (2007). The Fight for Jerusalem: Radical Islam, the West, and the Future of the Holy City. Washington, D.C.: Regnery Publishing.

Lev, Yaacov (1991). State and Society in Fatimid Egypt. Leiden; New York: E.J. Brill.

McMahon, Arthur .L. (1913). "Holy Sepulchre". Catholic Encyclopedia. New York: Robert Appleton Company.

From: 'Olave (St.) Mogwell Street, de Mugwellestrate Old Bethlehem Hospital',

A Dictionary of London (1918)

From: 'Hospitals: St Mary without Bishopsgate',

A History of the County of London: Volume 1:

London within the Bars, Westminster and Southwark (1909)

J Andrews, 'Hardly a Hospital, but a Charity for Pauper Lunatics'? Therapeutics at Bethlem in the Seventeenth and Eighteenth Centuries', in J Barry and C Jones (eds), Medicine and Charity

Before the Welfare State (London and New York: Routledge, 1991 J Andrews et al., The History of Bethlem (London and New York: Routledge, 1997)

A Scull, C MacKenzie and N Hervey, Masters of Bedlam: The Transformation of the Mad-Doctoring Trade (Princeton University Press, 1996)

P Allderidge, 'Management and Mismanagement at Bedlam, 1547-1633', in C Webster (ed.), Health, Medicine and Mortality in the Sixteenth Century (Cambridge University Press, 1979),

P Allderidge, 'Bedlam: Fact or Fantasy?', in R Porter et al (eds), Anatomy of Madness: essays in the history of psychiatry, vol. 1 (London; New York: Tavistock Publications, 1985)

www.ingramcontent.com/pod-product-compliance
Lightning Source LLC
Chambersburg PA
CBHW081055170526

45166CB00006B/2077